Applied Mathematics

Edited by Bruno Carpentieri

Published in London, United Kingdom

IntechOpen

Supporting open minds since 2005

Applied Mathematics
http://dx.doi.org/10.5772/intechopen.78131
Edited by Bruno Carpentieri

Contributors
Vicente Aboites, Miguel Ramirez, Franco Robledo - Pedeciba Informatica, Pablo Romero, Pablo Sartor, Snezana S. Djordjevic, Frank Xuyan Wang, Vakeel Ahmad Khan, Hira Fatima, Mobeen Ahmad, Ruma Kareem K. Ajeena

First published in London, United Kingdom, 2019 by IntechOpen
IntechOpen is the global imprint of INTECHOPEN LIMITED, registered in England and Wales, registration number: 11086078, The Shard, 25th floor, 32 London Bridge Street
London, SE19SG - United Kingdom
Printed in Croatia

British Library Cataloguing-in-Publication Data
A catalogue record for this book is available from the British Library

Additional hard and PDF copies can be obtained from orders@intechopen.com

Applied Mathematics
Edited by Bruno Carpentieri
p. cm.
Print ISBN 978-1-78923-905-8
Online ISBN 978-1-78923-906-5
eBook (PDF) ISBN 978-1-83962-222-9

For EU product safety concerns:
IN TECH d.o.o., Prolaz Marije Krucifikse Kozulić 3, 51000 Rijeka, Croatia, info@intechopen.com or visit our website at intechopen.com.

Meet the editor

Bruno Carpentieri obtained a Laurea degree in Applied Mathematics in 1997 from Bari University. He then furthered his PhD studies in Computer Science at the Institut National Polytechnique de Toulouse, France. After a postdoctoral appointment at the Institute of Mathematics and Scientific Computing, University of Graz, and as a consultant for a European project in cardiac modeling at CRS4 in Sardinia, Italy, he served as an assistant professor at the Bernoulli Institute for Mathematics, Computer Science and Artificial Intelligence, University of Groningen, and as a reader in Applied Mathematics at the School of Science and Technology, Nottingham Trent University in the UK. Since May 2017 he has held an associate professor appointment in applied mathematics at the Faculty of Computer Science, University of Bozen-Bolzano. His research interests are in the field of applied mathematics, numerical linear algebra, and high-performance computing.

Bruno Carpentieri served as a member of the scientific advisory board of several conference panels in computational mathematics and high-performance scientific computing (ENUMATH'07, Beteq'07, Beteq'08, Beteq'09, CEM'11, CEM'13, CEM'15, CEM'17, CEM'18, HPC2014, HPC2015, HPC2016, HPC2017, HPC2018, HPC2019, ICBCB 2017, and HPC/SmartTechCon2017). He is an editorial board member of the *Journal of Applied Mathematics*, an editorial committee member of *Mathematical Reviews* (American Mathematical Society), and a reviewer of about 30 scientific journals in numerical analysis and scientific computing. He has supervised 20 students' projects at BSc, MSc, and PhD level, and he is the author of about 40 publications in peer-reviewed scientific journals.

Contents

Preface **XI**

Chapter 1 **1**
Some Unconstrained Optimization Methods
by Snezana S. Djordjevic

Chapter 2 **29**
Unconstrained Optimization Methods: Conjugate Gradient Methods and Trust-Region Methods
by Snezana S. Djordjevic

Chapter 3 **47**
What Determines EP Curve Shape?
by Frank Xuyan Wang

Chapter 4 **73**
Topological Properties and Dynamic Programming Approach for Designing the Access Network
by Franco Robledo, Pablo Romero, Pablo Sartor, Luis Stábile and Omar Viera

Chapter 5 **89**
The Graphs for Elliptic Curve Cryptography
by Ruma Kareem K. Ajeena

Chapter 6 **111**
A Study of Bounded Variation Sequence Spaces
by Vakeel Ahmad Khan, Hira Fatima and Mobeen Ahmad

Chapter 7 **129**
Simple Approach to Special Polynomials: Laguerre, Hermite, Legendre, Tchebycheff, and Gegenbauer
by Vicente Aboites and Miguel Ramírez

Preface

This book contains well-written monographs within the broad spectrum of applied mathematics, with the aim of offering an interesting reading of current trends and problems in this fascinating and critically important field of mathematics to a broad category of researchers and practitioners. Recent developments in high-performance computing are radically changing the way we do numerics as applied mathematicians. Because of the impressive advances in computer technology and the introduction of fast methods that require less algorithmic cost and fewer memory resources, nowadays a rigorous numerical solution of many difficult computational science applications has become possible. In the future we will be solving much bigger problems, and even more factors will need to be considered than in the past when attempting to identify the optimal solution approach. The gap between fast and slow algorithms is rapidly growing. Methods that do more operations per grid node, cell, or element, such as higher-order and discontinuous Galerkin discretization schemes and spectral element methods, are becoming very attractive to use against more traditional techniques such as finite element discretization schemes. Structured data are already coming back, because they may achieve a better load balance than unstructured grids on computers with hundreds of thousands of processors. Novel classes of numerical methods with reduced computational complexity will need to be found to solve large-scale problems arising in an industrial setting.

The book is structured in three distinct parts, according to the aims and methodologies used by the authors in the development of their studies, ranging from optimization techniques to graph-oriented approaches and approximation theory, providing overall a good mix of both theory and practice. Chapters 1–2 present an overview of unconstrained optimization techniques, covering both line search and trust region methods that are essential ingredients to guarantee global convergence of descent schemes. Numerical optimization is the primary tool used in Chapter 3 to analyze the shape factor of exceedance probability curves, which is a critical analysis tool to assess risks, e.g., in the study of natural disasters such as floods, hurricanes, and earthquakes. Chapters 4–5 describe graph-oriented approaches. Chapter 4 develops a graph-based model for the topological design of the wide area network using dynamic programming and dynamic programming with state-space relaxation methodologies. Chapter 5 uses graph and subgraph models to speed up the computations of scalar multiplication algorithms on elliptic curves defined over finite fields, which is one central and time-consuming operation in elliptic curve cryptography. Finally, the contributions of the last two chapters deal with some aspects of functional approximation. Chapter 6 proposes a study of different forms of bounded variation sequence spaces of invariant means with the help of ideal operators and functions such as Orlicz function and modulus function. The results show the potential of the new theoretical tools to deal with the convergence problems of sequences in sigma-bounded variation occurring in many branches of science, engineering, and applied mathematics. Chapter 7 is devoted to an overview of the mathematics of special polynomials showing how to obtain them in a simple and straightforward approach using basic linear algebra concepts. Overall, the collection of contributions demonstrates the highly interdisciplinary character of the discipline, and emphasizes the continuing need for close cooperation between applied mathematicians, experimental physicists, engineers, and computer scientists in modern applied data science.

We express appreciation to all those who helped in the preparation of this book, and in particular to Luka Cvjetković at IntechOpen for his tireless editorship assistance.

Bruno Carpentieri
Free University of Bozen-Bolzano,
Bolzano, Italy

Chapter 1

Some Unconstrained Optimization Methods

Snezana S. Djordjevic

Abstract

Although it is a very old theme, unconstrained optimization is an area which is always actual for many scientists. Today, the results of unconstrained optimization are applied in different branches of science, as well as generally in practice. Here, we present the line search techniques. Further, in this chapter we consider some unconstrained optimization methods. We try to present these methods but also to present some contemporary results in this area.

Keywords: unconstrained optimization, line search, steepest descent method, Barzilai-Borwein method, Newton method, modified Newton method, inexact Newton method, quasi-Newton method

1. Introduction

Optimization is a very old subject of a great interest; we can search deep into a human history to find important examples of applying optimization in the usual life of a human being, for example, the need of finding the best way to produce food yielded finding the best piece of land for producing, as well as (later on, how the time was going) the best ways of treatment of the chosen land and the chosen seedlings to get the best results.

From the very beginning of manufacturing, the manufacturers were trying to find the ways to get maximum income with minimum expenses.

There are plenty of examples of optimization processes in pharmacology (for determination of the geometry of a molecule), in meteorology, in optimization of a trajectory of a deep-water vehicle, in optimization of power management (optimization of the production of electrical power plants), etc.

Optimization presents an important tool in decision theory and analysis of physical systems.

Optimization theory is a very developed area with its wide application in science, engineering, business management, military, and space technology.

Optimization can be defined as the process of finding the best solution to a problem in a certain sense and under certain conditions.

Along with the passage of time, optimization was evolving. Optimization became an independent area of mathematics in 1940, when Dantzig presented the so-called simplex algorithm for linear programming.

The development of nonlinear programming became great after presentation of conjugate gradient methods and quasi-Newton methods in the 1950s.

Today, there exist many modern optimization methods which are made to solve a variety of optimization problems. Now, they present the necessary tool for solving problems in diverse fields.

At the beginning, it is necessary to define an objective function, which, for example, could be a technical expense, profit or purity of materials, time, potential energy, etc.

The object function depends on certain characteristics of the system, which are known as variables. The goal is to find the values of those variables, for which the object function reaches its best value, which we call an extremum or an optimum.

It can happen that those variables are chosen in such a way that they satisfy certain conditions, i.e., restrictions.

The process of identifying the object function, variables, and restrictions for the given problem is called *modeling*.

The first and the most important step in an optimization process is the construction of the appropriate model, and this step can be the problem by itself. Namely, in the case that the model is too much simplified, it cannot be a faithful reflection of the practical problem. By the other side, if the constructed model is too complicated, then solving the problem is also too complicated.

After the construction of the appropriate model, it is necessary to apply the appropriate algorithm to solve the problem. It is no need to emphasize that there does not exist a universal algorithm for solving the set problem.

Sometimes, in the applications, the set of input parameters is bounded, i.e., the input parameters have values within the allowed space of input parameters D_x; we can write

$$x \in D_x. \tag{1}$$

Except (1), the next conditions can also be imposed:

$$\varphi_l(x_1, \dots, x_n) = \varphi_{0l}, l = 1, \dots, m_1 < n, \tag{2}$$

$$\psi_j(x_1, \dots, x_n) \le \psi_{0j}, j = 1, \dots, m_2. \tag{3}$$

Optimization task is to find the minimum (maximum) of the objective function $f(x) = f(x_1, \dots, x_n)$, under the conditions (1), (2), and (3).

If the object function is linear, and the functions $\varphi_l(x_1, \dots, x_n)\ l = 1, \dots, m_1$ and $\psi_j(x_1, \dots, x_n) j = 1, \dots, m_2$ are linear, then it is about the linear programming problem, but if at least one of the mentioned functions is nonlinear, it is about the nonlinear programming problem.

Unconstrained optimization problem can be presented as

$$\min_{x \in R^n} f(x), \tag{4}$$

where $f \in R^n$ is a smooth function.

Problem (4) is, in fact, the unconstrained minimization problem. But, it is well known that the unconstrained minimization problem is equivalent to an unconstrained maximization problem, i.e.

$$\min f(x) = -\max(-f(x)), \tag{5}$$

as well as

$$\max f(x) = -\min(-f(x)). \tag{6}$$

Definition 1.1.1 x^* *is called a global minimizer of* f *iff* $f(x^*) \leq f(x)$ *for all* $x \in R^n$.

The ideal situation is finding a global minimizer of f. Because of the fact that our knowledge of the function f is usually only local, the global minimizer can be very difficult to find. We usually do not have the total knowledge about f. In fact, most algorithms are able to find only a local minimizer, i.e., a point that achieves the smallest value of f in its neighborhood.

So, we could be satisfied by finding the local minimizer of the function f. We distinguish weak and strict (or strong) local minimizer.

Formal definitions of local weak and strict minimizer of the function f are the next two definitions, respectively.

Definition 1.1.2 x^* *is called a weak local minimizer of* f *if there exists a neighborhood* N *of* x^**, such that* $f(x^*) \leq f(x)$ *for all* $x \in N$.

Definition 1.1.3 x^* *is called a strict (strong) local minimizer of* f *if there exists a neighborhood* N *of* x^**, such that* $f(x^*) < f(x)$ *for all* $x \in N$.

Considering backward definitions 1.1.2 and 1.1.3, the procedure of finding local minimizer (weak or strict) does not seem such easy; it seems that we should examine all points from the neighborhood of x^*, and it looks like a very difficult task.

Fortunately, if the object function f satisfies some special conditions, we can solve this task in a much easier way.

For example, we can assume that the object function f is smooth or, furthermore, twice continuously differentiable. Then, we concentrate to the gradient $\nabla f(x^*)$ as well as to the Hessian $\nabla^2 f(x^*)$.

All algorithms for unconstrained minimization require the user to start from a certain point, so-called the starting point, which we usually denote by x_0. It is good to choose x_0 such that it is a reasonable estimation of the solution. But, to find such estimation, a little more knowledge about the considered set of data is needed, and the systematic investigation is needed also. So, it seems much simpler to use one of the algorithms to find x_0 or to take it arbitrarily.

There exist two important classes of iterative methods—*line search methods* and *trust-region methods*—made in the aim to solve the unconstrained optimization problem (4).

In this chapter, at first, we discuss different kinds of line search. Then, we consider some line search optimization methods in details, i.e., we study steepest descent method, Barzilai-Borwein gradient method, Newton method, and quasi-Newton method.

Also, we try to give some of the most recent results in these areas.

2. Line search

Now, let us consider the problem

$$\min_{x \in R^n} f(x), \tag{7}$$

where $f : R^n \to R$ is a continuously differentiable function, bounded from below.

There exists a great number of methods made in the aim to solve the problem (7).

The optimization methods based on line search utilize the next iterative scheme:

$$x_{k+1} = x_k + t_k d_k, \tag{8}$$

where x_k is the current iterative point, x_{k+1} is the next iterative point, d_k is the search direction, and t_k is the step size in the direction d_k.

At first, we consider the monotone line search.

Now, we give the iterative scheme of this kind of search.

Algorithm 1.2.1. *(Monotone line search).*
Assumptions: $\epsilon > 0$, x_0, $k := 0$.
Step 1. If $\|g_k\| \le \epsilon$, then STOP.
Step 2. Find the descent direction d_k.
Step 3. Find the step size t_k, such that $f(x_k + t_k d_k) < f(x_k)$.
Step 4. Set $x_{k+1} = x_k + t_k d_k$.
Step 5. Take $k := k + 1$ and go to Step 1.
Denote

$$\Phi(t) = f(x_k + td_k).$$

Trying to solve the minimization problem, we are going to search for the step size $t = t_k$, in the direction d_k, such that the next relation holds:

$$\Phi(t_k) < \Phi(0).$$

That procedure is called the monotone line search.

We can search for the step size t_k in such a way that the next relation holds:

$$f(x_k + t_k d_k) = \min_{t \ge 0} f(x_k + t_k d_k), \tag{9}$$

i.e.

$$\Phi(t_k) = \min_{t \ge 0} \Phi(t), \tag{10}$$

or we can use the next formula:

$$t_k = \min\left\{t \,|\, g(x_k + td_k)^T d_k = 0, t \ge 0\right\}. \tag{11}$$

In this case we are talking about *the exact* or *the optimal* line search, where the parameter t_k, which is received as the solution of the one-dimensional problem (10), is *the optimal step size*.

By the other side, instead of using the relation (9), or the relation (11), we can be satisfied by searching for such t_k, which is acceptable if the next relation suits us:

$$f(x_k) - f(x_k + t_k d_k) > \delta_k > 0.$$

Then, we are talking about *the inexact* or *the approximate* or *the acceptable* line search, which is very much utilized in the practice.

There are several reasons to use the inexact instead of the exact line search. One of them is that the exact line search is expensive. Further, in the cases when the iteration is far from the solution, the exact line search is not efficient. Next, in the practice, the convergence rate of many optimization methods (such as Newton or quasi-Newton) does not depend on the exact line search.

First, we are going to mention so-called basic and, by the way, very well-known inexact line searches.

Algorithm 1.2.2. *(Backtracking).*
Assumptions: x_k, the descent direction d_k, $0 < \delta < \frac{1}{2}$, $\eta \in (0, 1)$.
Step 1. $t := 1$.

Step 2. While $f(x_k + td_k) > f(x_k) + \delta t g_k^T d_k$, $t := t \cdot \eta$.
Step 3. Set $t_k = t$.
Now, we describe the Armijo rule.

Theorem 1.2.1. [1] *Let $f \in C^1(R^n)$ and let d_k be the descent direction. Then, there exists the nonnegative number m_k, such that*

$$f(x_k + \eta^{m_k} d_k) \le f(x_k) + c_1 \eta^{m_k} g_k^T d_k,$$

where $c_1 \in (0, 1)$ and $\eta \in (0, 1)$.
Next, we describe the Goldstein rule [2].
The step size t_k is chosen in such a way that

$$\begin{aligned} f(x_k + td_k) &\le f(x_k) + \delta t g_k^T d_k, \\ f(x_k + td_k) &> f(x_k) + (1 - \delta) t g_k^T d_k, \end{aligned}$$

where $0 < \delta < \frac{1}{2}$.
Now, Wolfe line search rules follow [3], [4].
Standard Wolfe line search conditions are

$$f(x_k + t_k d_k) - f(x_k) \le \delta t_k g_k^T d_k, \tag{12}$$

$$g_{k+1}^T d_k \ge \sigma g_k^T d_k, \tag{13}$$

where d_k is a descent direction and $0 < \delta \le \sigma < 1$.

This efficient strategy means that we should accept a positive step length t_k, if conditions (12)–(13) are satisfied.

Strong Wolfe line search conditions consist of (12) and the next, stronger version of (13):

$$|g_{k+1}^T d_k| \le -\sigma g_k^T d_k. \tag{14}$$

In the generalized Wolfe line search conditions, the absolute value in (14) is replaced by the inequalities:

$$\sigma_1 g_k^T d_k \le g_{k+1}^T d_k \le -\sigma_2 g_k^T d_k, \; 0 < \delta \le \sigma_1 < 1, \; \sigma_2 \ge 0. \tag{15}$$

By the other side, in the approximate Wolfe line search conditions, the inequalities (15) are changed into the next ones:

$$\sigma g_k^T d_k \le g_{k+1}^T d_k \le (2\delta - 1) g_k^T d_k, \; 0 < \delta < \frac{1}{2}, \; \delta < \sigma < 1. \tag{16}$$

The next lemma is very important.

Lemma 1.2.1. [5] *Let $f \in C(R^n)$. Let d_k be a descent direction at the point x_k, and assume that the function f is bounded from below along the direction $\{x_k + td_k | t > 0\}$. Then, if $0 < \delta < \sigma < 1$, there exist the intervals inside which the step length satisfies standard Wolfe conditions and strong Wolfe conditions.*

By the other side, the introduction of the non-monotone line search is motivated by the existence of the problems where the search direction does not have to be a descent direction. This can happen, for example, in stochastic optimization [6].

Next, some efficient quasi-Newton methods, for example, *SR*1 update, do not produce the descent direction in every iteration [5].

Further, some efficient methods like spectral are not monotone at all.

Some numerical results given in [7–11] show that non-monotone techniques are better than the monotone ones if the problem is to find the global optimal values of the object function.

Algorithms of the non-monotone line search do not insist on a descent of the object function in every step. But, even these algorithms require the reduction of the object function after a predetermined number of iterations.

The first non-monotone line search technique is presented in [12]. Namely, in [12], the problem is to find the step size which satisfies

$$f(x_k + t_k d_k) \leq \max_{0 \leq j \leq m(k)} f(x_{k-j}) + \delta t_k g_k^T d_k,$$

where $m(0) = 0$, $0 \leq m(k) \leq \min\{m(k-1)+1, M\}$, for $k \geq 1$, $\delta \in (0,1)$, where M is a nonnegative integer.

This strategy is in fact the generalization of Armijo line search. In the same work, the authors suppose that the search directions satisfy the next conditions for some positive constants b_1 and b_2:

$$g_k^T d_k \leq -b_1 \|g_k\|^2,$$
$$\|d_k\| \leq b_2 \|g_k\|.$$

The next non-monotone line search is described in [11].

Let x_0 be the starting point, and let

$$0 \leq \eta_{min} \leq \eta_{max} \leq 1,\ 0 < \delta < \sigma < 1 < \rho,\ \mu > 0.$$

Let $C_0 = f(x_0)$, $Q_0 = 1$.

The step size has to satisfy the next conditions:

$$f(x_k + t_k d_k) \leq C_k + \delta t_k g_k^T d_k, \tag{17}$$

$$g(x_k + t_k d_k) \geq \sigma g_k^T d_k. \tag{18}$$

The value η_k is chosen from the interval $[\eta_{min}, \eta_{max}]$ and then

$$Q_{k+1} = \eta_k Q_k + 1,\ C_{k+1} = \frac{\eta_k Q_k C_k + f(x_{k+1})}{Q_{k+1}}.$$

Non-monotone rules which contain the sequence of nonnegative parameters $\{\epsilon_k\}$ are used firstly in [13], and they are successfully used in many other algorithms, for example, in [14]. The next property of the parameters ϵ_k is assumed:

$$\epsilon_k > 0,\quad \sum_k \epsilon_k = \epsilon < \infty,$$

and the corresponding rule is

$$f(x_k + t_k d_k) \leq f(x_k) + c_1 t_k g_k^T d_k + \epsilon_k.$$

Now, we give the non-monotone line search algorithm, shortly *NLSA*, presented in [11].

Algorithm 1.2.3. *(NLSA).*

Assumptions: x_0, $0 \leq \eta_{min} \leq \eta_{max} \leq 1$, $0 < \delta < \sigma < 1 < \rho$, $\mu > 0$.

Set $C_0 = f(x_0)$, $Q_0 = 1$, $k = 0$.

Step 1. If $\|\nabla f(x_k)\|$ is sufficiently small, then STOP.

Step 2. Set $x_{k+1} = x_k + t_k d_k$, where t_k satisfies either the (non-monotone) Wolfe conditions (17) and (18) or the (non-monotone) Armijo conditions: $t_k = \bar{t}_k \rho^{h_k}$, where $\bar{t}_k > 0$ is the trial step and h_k is the largest integer such that (17) holds and $t_k \leq \mu$.

Step 3. Choose $\eta_k \in [\eta_{min}, \eta_{max}]$, and set

$$Q_{k+1} = \eta_k Q_k + 1, C_{k+1} = (\eta_k Q_k C_k + f(x_{k+1}))/Q_{k+1}.$$

Step 4. Set $k := k + 1$ and go to Step 1.

We can notice [11] that C_{k+1} is a convex combination of $f(x_0), f(x_1), \dots, f(x_k)$. The parameter η_k controls the degree of non-monotonicity.

If $\eta_k = 0$ for all k, then this non-monotone line search becomes monotone Wolfe or Armijo line search.

If $\eta_k = 1$ for all k, then $C_k = A_k$, where

$$A_k = \frac{1}{k+1} \sum_{i=0}^{k} f(x_i).$$

Lemma 1.2.2. [11] *If $\nabla f(x_k)^T d_k \leq 0$ for each k, then for the iterates generated by the non-monotone line search algorithm, we have $f_k \leq C_k \leq A_k$ for each k. Moreover, if $\nabla f(x_k)^T d_k < 0$ and $f(x)$ are bounded from below, then there exists t_k satisfying either Wolfe or Armijo conditions of the line search update.*

This study would be very incomplete unless we mention that there are many modifications of the abovementioned line searches. All these modifications are made to improve the previous results.

For example, in [15], the new inexact line search is described by the next way.

Let $\beta \in (0, 1)$, $\sigma \in \left(0, \frac{1}{2}\right)$; let B_k be a symmetric positive definite matrix which approximates $\nabla^2 f(x_k)$ and $s_k = -\frac{g_k^T d_k}{d_k^T B_k d_k}$. The step size t_k is the largest one in $\{s_k, s_k\beta, s_k\beta^2, \dots\}$ such that

$$f(x_k + td_k) - f(x_k) \leq \sigma t \left[g_k^T d_k + \frac{1}{2} t d_k^T B_k d_k \right].$$

Further, in [16], a new inexact line search rule is presented. This rule is a modified version of the classical Armijo line search rule. We describe it now.

Let $g = \nabla f(x)$ be a Lipschitz continuous function and L the Lipschitz constant. Let L_k be an approximation of L. Set

$$\beta_k = -\frac{g_k^T d_k}{L_k \|d_k\|^2}.$$

Find a step size t_k as the largest component in the set $\{\beta_k, \beta_k\rho, \beta_k\rho^2 \dots\}$ such that the inequality

$$f(x_k + t_k d_k) \leq f(x_k) + \sigma t_k \left(g_k^T d_k - \frac{1}{2} t_k \mu L_k \|d_k\|^2 \right)$$

holds, where $\sigma \in (0, 1)$, $\mu \in [0, \infty)$, and $\rho \in (0, 1)$ are given constants.

Next, in [17], a new, modified Wolfe line search is given in the next way.

Find $t_k > 0$ such that

$$f(x_k + t_k d_k) - f(x_k) \leq \min\{\delta t_k g_k^T d_k, -\gamma t_k^2 \|d_k\|^2\},$$
$$g(x_k + t_k d_k)^T d_k \geq \sigma g_k^T d_k,$$

where $\delta \in (0,1)$, $\sigma \in (\delta, 1)$, and $\gamma > 0$.
More recent results on this topic can be found, for example, in [18–23].

2.1 Steepest descent (*SD*)

The classical steepest descent method which is designed by Cauchy [24] can be considered as one among the most important procedures for minimization of real-valued function defined on $\mathbb{R}^n$.

Steepest descent is one of the simplest minimization methods for unconstrained optimization. Since it uses the negative gradient as its search direction, it is known also as the gradient method.

It has low computational cost and low matrix storage requirement, because it does not need the computations of the second derivatives to be solved to calculate the search direction [25].

Suppose that $f(x)$ is continuously differentiable in a certain neighborhood of a point x_k and also suppose that $g_k \triangleq \nabla f(x_k) \neq 0$.

Using Taylor expansion of the function f near x_k as well as Cauchy-Schwartz inequality, one can easily prove that the greatest fall of f exists if and only if $d_k = -g_k$, i.e., $-g_k$ is the steepest descent direction.

The iterative scheme of the *SD* method is

$$x_{k+1} = x_k - t_k g_k. \tag{19}$$

The classical steepest descent method uses the exact line search.

Now, we give the algorithm of the steepest descent method which refers to the exact as well as to the inexact line search.

Algorithm 1.2.4. *(Steepest descent method, i.e., SD method).*
Assumptions: $0 < \epsilon \ll 1$, $x_0 \in \mathbb{R}^n$. Let $k = 0$.
Step 1. If $\|g_k\| \leq \epsilon$, then STOP, else set $d_k = -g_k$.
Step 2. Find the step size t_k, which is the solution of the problem

$$\min_{t \geq 0} f(x_k + t d_k), \tag{20}$$

else find the step size t_k by any of the inexact line search methods.
Step 3. Set $x_{k+1} = x_k + t_k d_k$.
Step 4. Set $k := k + 1$ and go to Step 1.

The classical and the oldest steepest descent step size t_k, which was designed by Cauchy (in the case of the exact line search), is computed as [26]

$$t_k = \frac{g_k^T g_k}{g_k^T G g_k},$$

where $g_k = \nabla f(x_k)$ and $G = \nabla^2 f(x_k)$.

Theorem 1.2.2. [27] *(Global convergence theorem of the SD method) Let $f \in C^1$. Then, each accumulation point of the iterative sequence $\{x_k\}$, generated by Algorithm 1.2.4, is a stationary point.*

Remark 1.2.1. *The steepest descent method has at least the linear convergence rate.*

More information about the convergence of the *SD* method can be found in [5, 27].

Although known as the first unconstrained optimization method, this method is still a theme considered by scientists.

Different modifications of this method are made, for example, see [25, 28–32].

In [28], the authors presented a new search direction from Cauchy's method in the form of two parameters known as *Zubai'ah-Mustafa-Rivaie-Ismail* method, shortly, *ZMRI* method:

$$d_k = -g_k - \|g_k\| g_{k-1}. \tag{21}$$

So, in [28], a new modification of *SD* method is suggested using a new search direction, d_k, given by (21). The numerical results are presented based on the number of iterations and CPU time. It is shown that this new method is efficient when it is compared to the classical *SD*.

In [25], a new scaled search direction of *SD* method is presented. The inspiration for this new method is the work of Andrei [33], in which the author presents and analyzes a new scaled conjugate gradient algorithm, based on an interpretation of the secant equation and on the inexact Wolfe line search conditions.

The method proposed in [25] is known as *Rashidah-Rivaie-Mamat* (*RRM*) method, and it suggests the direction d_k given by the next relation:

$$d_k = \begin{cases} -g_k, \text{ if } k = 0, \\ -\theta_k g_k - \|g_k\| g_{k-1}, \end{cases} \tag{22}$$

where θ_k is a scaling parameter, $\theta_k = \frac{d_{k-1}^T y_{k-1}}{\|g_{k-1}\|^2}$, $y_{k-1} = g_k - g_{k-1}$.

Further, in [25], a comparison among *RRM*, *ZMRI*, and *SD* methods is made; it is shown that *RRM* method is better than *ZMRI* and *SD* methods.

It is interesting that the exact line search is used in [25].

In [34], the properties of steepest descent method from the literature are reviewed together with advantages and disadvantages of each step size procedure.

Namely, the step size procedures, which are compared in this paper, are:

1. $t_k = \frac{g_k^T g_k}{g_k^T H_k g_k}$: Step size method by Cauchy [24], computed by exact line search (*C* step size).

2. Given $s > 0$, $\beta, \sigma \in (0,1)$, $t_k = \max\{s, s\beta, s\beta^2, \ldots\}$ such that

$$f(x_k + t_k d_k) \le f(x_k) + \sigma t_k g_k^T d_k - \text{Armijo's line search (A step size)}.$$

3. Given $\beta, \sigma \in (0,1)$, $\tilde{t}_0 = 1$, and $t_k = \beta \tilde{t}_k$ such that

$$f(x_k + t_k d_k) \le f(x_k) + \sigma t_k g_k^T d_k - \text{Backtracking line search (B step size)}.$$

4. $t_k = \frac{s_{k-1}^T y_{k-1}}{\|y_{k-1}\|^2}$, (*BB*1), $t_k = \frac{\|s_{k-1}\|^2}{s_{k-1}^T y_{k-1}}$, (*BB*2), $s_{k-1} = x_k - x_{k-1}$ $y_{k-1} = g_k - g_{k-1}$: Barzilai and Borwein's formula. The convergence is R-superlinear.

5. $t_k = \frac{t_{k-1}^2 g_k^T g_k}{2(f(x_k + t_k d_k) - f(x_k) + t_{k-1} g_k^T g_k)}$: Elimination line search (*EL* step size), which estimates the step size without computation of the Hessian.

The comparison is based on time execution, number of total iteration, total percentage of function, gradient and Hessian evaluation, and the most decreased value of objective function obtained.

From the numerical results, the authors conclude that the A method and $BB1$ method are the best methods among others.

Further, in [34], the general conclusions about the steepest descent method are given:

1. This method is sensitive to the initial point.

2. This method has a descent property, and it is a logical starting procedure for all gradient based methods.

3. x_k approaches the minimizer slowly, in fact in a zigzag way.

In [35], in the aim to achieve fast convergence and the monotone property, a new step size for the steepest descent method is suggested.

In [36], for quadratic positive definite problems, an over-relaxation has been considered. Namely, Raydan and Svaiter [36] proved that the poor behavior of the steepest descent method is due to the optimal Cauchy choice of step size and not to the choice of the search direction. These results are extended in [29] to convex, well-conditioned functions. Further, in [29], it is shown that a simple modification of the step length by means of a random variable uniformly distributed in $(0, 1]$, for the strongly convex functions, represents an improvement of the classical gradient descent algorithm. Namely, in this paper, the idea is to modify the gradient descent method by introducing a relaxation of the following form:

$$x_{k+1} = x_k + \theta_k t_k d_k, \tag{23}$$

where θ_k is the relaxation parameter, a random variable uniformly distributed between 0 and 1.

In the recent years, the steepest descent method has been applied in many branches of science; one can be inspired, for example, by [37–43].

2.2 Barzilai and Borwein gradient method

Remind to the fact that SD method performs poorly, converges linearly, and is badly affected by the ill-conditioning.

Also, remind to the fact that this poor behavior of SD method is due to the optimal choice of the step size and not to the choice of the steepest descent direction $-g_k$.

Barzilai and Borwein presented [44] a two-point step size gradient method, which is well known as BB method.

The step size is derived from a two-point approximation to the secant equation.

Consider the gradient iteration form:

$$x_{k+1} = x_k - t_k g_k.$$

It can be rewritten as $x_{k+1} = x_k - D_k g_k$, where $D_k = t_k I$.

To make the matrix D_k having quasi-Newton property, the step size t_k is computed in such a way that we get

$$\min \| s_{k-1} - D_k y_{k-1} \|.$$

This yields that

$$t_k^{BB1} = \frac{s_{k-1}^T y_{k-1}}{y_{k-1}^T y_{k-1}}, s_{k-1} = x_k - x_{k-1}, y_{k-1} = g_k - g_{k-1}. \tag{24}$$

But, using symmetry, we may minimize $\|D_k^{-1} s_{k-1} - y_{k-1}\|$, with respect to t_k, and we get:

$$t_k^{BB2} = \frac{\|s_{k-1}\|^2}{s_{k-1}^T y_{k-1}}, s_{k-1} = x_k - x_{k-1}, y_{k-1} = g_k - g_{k-1}. \tag{25}$$

Now, we give the algorithm of *BB* method.

Algorithm 1.2.5. *(Barzilai-Borwein gradient method, i.e., BB method).*
Assumptions: $0 < \epsilon \ll 1$, $x_0 \in \mathbb{R}^n$. Let $k = 0$.
Step 1. If $\|g_k\| \leq \epsilon$, then STOP, else set $d_k = -g_k$.
Step 2. If $k = 0$, then find the step size t_0 by the line search, else compute t_k using the formula (24) or (25).
Step 3. Set $x_{k+1} = x_k + t_k d_k$.
Step 4. Set $k := k + 1$ and go to Step 1.
Considering Algorithm 1.2.5, we can conclude that this method does not require any matrix computation or any line search.

The Barzilai-Borwein method is in fact the gradient method, which requires less computational work than *SD* method, and it speeds up the convergence of the gradient method. Barzilai and Borwein proved that *BB* algorithm is $R-$superlinearly convergent for the quadratic case.

In the general non-quadratic case, a globalization strategy based on non-monotone line search is applied in this method.

In this general case, t_k, computed by (24) or (25), may be unacceptably large or small. That is the reason why we assume that there exist the numbers t^l and t^r, such that

$$0 < t^l \leq t_k \leq t^r, \text{ for all } k.$$

Using the iteration

$$x_{k+1} = x_k - \frac{1}{t_k} g_k = x_k - \lambda_k g_k, \tag{26}$$

with

$$t_k = \frac{s_{k-1}^T y_{k-1}}{s_{k-1}^T s_{k-1}}, \lambda_k = \frac{1}{t_k},$$

$$s_k = -\frac{1}{t_k} g_k = -\lambda_k g_k,$$

we get

$$t_{k+1} = \frac{s_k^T y_k}{s_k^T s_k} = \frac{-\lambda_k g_k^T y_k}{\lambda_k^2 g_k^T g_k} = -\frac{g_k^T y_k}{\lambda_k g_k^T g_k}.$$

Now, we give the algorithm of the Barzilai-Borwein method with non-monotone line search.

Algorithm 1.2.6. *(BB method with non-monotone line search).*

Assumptions: $0 < \epsilon \ll 1$, $x_0 \in \mathbb{R}^n$, $M \geq 0$ is an integer, $\rho \in (0,1)$, $\delta > 0$, $0 < \sigma_1 < \sigma_2 < 1$, t^l, t^r. Let $k = 0$.

Step 1. If $\|g_k\| \leq \epsilon$, then STOP.

Step 2. If $t_k \leq t^l$, or $t_k \geq t^r$, then set $t_k = \delta$.

Step 3. Set $\lambda = \frac{1}{t_k}$.

Step 4. (non-monotone line search) If

$$f(x_k - \lambda g_k) \leq \max_{0 \leq j \leq \min(k,M)} f(x_{k-j}) - \rho\lambda g_k^T g_k,$$

then set

$$\lambda_k = \lambda, x_{k+1} = x_k - \lambda_k g_k,$$

and go to Step 6.

Step 5. Choose $\sigma \in [\sigma_1, \sigma_2]$, set $\lambda = \sigma\lambda$, and go to Step 4.

Step 6. Set $t_{k+1} = -\frac{g_k^T y_k}{\lambda_k g_k^T g_k}$ and $k := k + 1$, and return to Step 1.

Obviously, the above algorithm is globally convergent.

Several authors paid attention to the Barzilai-Borwein method, and they proposed some variants of this method.

In [8], the globally convergent Barzilai-Borwein method is proposed by using non-monotone line search by Grippo et al. [12]. In the same paper, Raydan proves the global convergence of the non-monotone Barzilai-Borwein method.

Further, Grippo and Sciandrone [45] propose another type of the non-monotone Barzilai-Borwein method.

Dai [7] gives the basic analysis of the non-monotone line search strategy.

Moreover, in [46] numerical results are presented, using

$$t_k = \frac{s_{\nu(k)}^T y_{\nu(k)}}{s_{\nu(k)}^T s_{\nu(k)}}. \tag{27}$$

and

$$\nu(k) = M_c \cdot \left\lfloor \frac{k-1}{M_c} \right\rfloor,$$

where for $r \in \mathbb{R}$, $\lfloor r \rfloor$ denotes the largest integer j such that $j \leq r$ and Mc is a positive integer. The gradient method with (27) is called the cyclic Barzilai-Borwein method. Numerical results in [46] prove that their method performs better than the Barzilai-Borwein method.

Many researchers study the gradient method for minimizing a strictly convex quadratic function, namely,

$$\min f(x) = \frac{1}{2} x^T A x - b^T x, \tag{28}$$

where $A \in \mathbb{R}^{n \times n}$ is a symmetric positive definite matrix and $b \in \mathbb{R}^n$ is a given vector. For an application of the Barzilai-Borwein method to the problem (28), Raydan [47] establishes global convergence, and Dai and Liao [48] prove $\mathbb{R}$-linear rate of convergence. Friedlander, Martinez, Molina, and Raydan [49] propose a new gradient method with retards, in which t_k is defined by

$$t_k = \frac{g_{\nu(k)}^T A^{\rho(k)+1} g_{\nu(k)}}{g_{\nu(k)}^T A^{\rho(k)} g_{\nu(k)}}, \nu(k) \in \{k, k-1, \ldots, \max\{0, k-m\}\} \tag{29}$$

and $\rho(k) \in \{q_1, \ldots, q_m\}$, where m is a positive integer and $q_1, \ldots, q_m \geq -2$ are integers. In the same paper, they establish its global convergence for problem (28) and prove the Q-superlinear rate of convergence in the special case.

In [50], the authors extend the Barzilai-Borwein method, and they give *extended Barzilai-Borwein method*, which they denote *EBB*. They also establish global and $Q-$superlinear convergence properties of the proposed method for minimizing a strictly convex quadratic function. Furthermore, they discuss an application of their method to general objective functions. In [50], a new step size is proposed by extending (29). Namely, in this paper, following Friedlander et al. [49], a new step size is proposed as follows:

$$t_k = \sum_{i=1}^{l} \phi_i \frac{g_{\nu_{i(k)}}^T A^{\rho_i(k)+1} g_{\nu_i(k)}}{g_{\nu_{i(k)}}^T A^{\rho_i(k)} g_{\nu_i(k)}},$$

$$\phi_i \geq 0, \sum_{i=1}^{n} \phi_i = 1,$$

$$\nu_i(k) \in \{k, k-1, \ldots, \max\{0, k-m\}\}$$

and

$$\phi_i(k) \in \{q_1, \ldots, q_m\},$$

where l and m are positive integers and $q_1, \ldots, q_m$ are integers.

Also, an application of algorithm *EBB* to general unconstrained minimization problems (4) is considered.

Following Raydan [8], the authors [50] further combine the non-monotone line search and algorithm *EBB* to get the algorithm called *NEBB*. They also prove the global convergence of the algorithm *NEBB*, under some classical assumptions.

The Barzilai-Borwein method and its related methods are reviewed by Dai and Yuan [51] and Fletcher [52].

In [53], a new concept of the approximate optimal step size for gradient method is introduced and used to interpret the *BB* method; an efficient gradient method with the approximate optimal step size for unconstrained optimization is presented. The next definition is introduced in [53].

Definition 1.2.1. *Let* $\Phi(t)$ *be an approximation model of* $f(x_k - tg_k)$. *A positive constant* t^* *is called approximate optimal step size associated to* $\Phi(t)$ *for gradient method, if* t^* *satisfies*

$$t^* = \arg\min_{t>0} \Phi(t).$$

The approximate optimal step size is different from the steepest descent step size, which will lead to the expensive computational cost. The approximate optimal step size is generally calculated easily, and it can be applied to unconstrained optimization.

Due to the effectiveness of t_k^{BB1} and the fact that $t_k^{BB1} = \arg\min_{t>0}\Phi(t)$, we can naturally ask if more suitable approximation models can be constructed to generate more efficient approximate optimal step-sizes.

This is the purpose of work [53]. Further, if the objective function $f(x)$ is not close to a quadratic function on the line segment between x_{k-1} and x_k, in this paper a conic model is developed to generate the approximate optimal step size if the conic model is suitable to be used. Otherwise, the authors consider two cases:

i. If $s_{k-1}^T y_{k-1} > 0$, the authors construct a new quadratic model, to derive the approximate optimal step size.

ii. If $s_{k-1}^T y_{k-1} \leq 0$, they construct a new quadratic model or two other new approximation models to generate the approximate optimal step size for gradient method. They also analyze the convergence of the proposed method under some suitable conditions. Numerical results show the proposed method is better than the BB method.

In [54], derivative-free iterative scheme that uses the residual vector as search direction for solving large-scale systems of nonlinear monotone equations is presented.

The Barzilai-Borwein method is widely used; some interesting results can be found in [55–57].

2.3 Newton method

The basic idea of Newton method for unconstrained optimization is the iterative usage of the quadratic approximation $q^{(k)}$ to the objective function f at the current iterate x_k and then minimization of such approximation $q^{(k)}$.

Let $f: \mathbb{R}^n \to \mathbb{R}$ be twice continuously differentiable, $x_k \in \mathbb{R}^n$, and let the Hessian $\nabla^2 f(x_k)$ be positive definite.

We model f at the current point x_k by the quadratic approximation $q^{(k)}$:

$$f(x_k + s) \approx q^{(k)}(s) = f(x_k) + \nabla f(x_k)^T s + \frac{1}{2} s^T \nabla^2 f(x_k) s, \; s = x - x_k.$$

Minimization of $q^{(k)}(s)$ gives the next iterative scheme:

$$x_{k+1} = x_k - \left(\nabla^2 f(x_k)\right)^{-1} \nabla f(x_k),$$

which is known as Newton formula.

Denote $G_k = \nabla^2 f(x_k), g_k = \nabla f(x_k)$.

Then, we have a simpler form:

$$x_{k+1} = x_k - G_k^{-1} g_k. \tag{30}$$

A *Newton direction* is

$$s_k = x_{k+1} - x_k = -G_k^{-1} g_k. \tag{31}$$

We have supposed that G_k is positive definite. So, the Newton direction is a descent direction. This we can conclude from

$$g_k^T s_k = -g_k^T G_k^{-1} g_k < 0.$$

Now, we give the algorithm of the Newton method.

Algorithm 1.2.7. *(Newton method).*
Assumptions: $\epsilon > 0$, $x_0 \in \mathbb{R}^n$. Let $k = 0$.
Step 1. If $\|g_k\| \leq \epsilon$, then STOP.
Step 2. Solve $G_k s = -g_k$ for s_k.
Step 3. Set $x_{k+1} = x_k + s_k$.
Step 4. $k := k + 1$, return to Step 1.
The next theorem shows the local convergence and the quadratic convergence rate of Newton method.

Theorem 1.2.3. [27] *(Convergence theorem of Newton method) Let $f \in C^2$ and x_k be close enough to the solution x^* of the minimization problem with $g(x^*) = 0$. If the Hessian $G(x^*)$ is positively definite and $G(x)$ satisfies Lipschitz condition*

$$|G_{ij}(x) - G_{ij}(y)| \leq \beta \|x - y\|, \textit{for some } \beta, \textit{for all } i, j,$$

where $G_{ij}(x)$ is the (i,j) element of $G(x)$ and then for all k, Newton direction (31) is well-defined; the generated sequence $\{x_k\}$ converges to x^* with a quadratic rate.

But, in spite of this quadratic rate, the Newton method is a local method: when the starting point is far away from the solution, there is a possibility that G_k is not positive definite, as well as Newton direction is not a descent direction.

So, to guarantee the global convergence, we can use Newton method with line search. We can remind to the fact that only when the step size sequence $\{t_k\}$ tends to 1, Newton method is convergent with the quadratic rate.

Newton iteration with line search is as follows:

$$d_k = -G_k^{-1} g_k, \tag{32}$$

$$x_{k+1} = x_k + t_k d_k. \tag{33}$$

Now, we give the algorithm.

Algorithm 1.2.8. *(Newton method with line search).*
Assumptions: $\epsilon > 0$, $x_0 \in \mathbb{R}^n$. Let $k = 0$.
Step 1. If $\|g_k\| \leq \epsilon$, then STOP.
Step 2. Solve $G_k d = -g_k$ for d_k.
Step 3. Line search step: find t_k such that

$$f(x_k + t_k d_k) = \min_{t \geq 0} f(x_k + t d_k),$$

or find t_k such that (inexact) Wolfe line search rules hold.
Step 4. Set $x_{k+1} = x_k + t_k d_k$ and $k = k + 1$, and go to Step 1.
The next theorems claim that Algorithm 1.2.8 with the exact line search, as well as Algorithm 1.2.8 with the inexact line search, are globally convergent.

Theorem 1.2.4. [27] *Let $f : \mathbb{R}^n \to \mathbb{R}$ be twice continuously differentiable on open convex set $D \subset \mathbb{R}^n$. Assume that for any $x_0 \in D$ there exists a constant $m > 0$, such that $f(x)$ satisfies*

$$u^T \nabla^2 f(x) u \geq m \|u\|^2, \textit{for all } u \in \mathbb{R}^n, x \in L(x_0), \tag{34}$$

where $L(x_0) = \{x | f(x) \leq f(x_0)\}$ is the corresponding level set. Then, the sequence $\{x_k\}$, generated by Algorithm 1.2.8, with the exact line search, satisfies:

1. When $\{x_k\}$ is a finite sequence, $g_k = 0$ for some k.

2. When $\{x_k\}$ is an infinite sequence, $\{x_k\}$ converges to the unique minimizer x^* of f.

Note that the next relation holds from the standard Wolfe line search:

$$f(x_k) - f(x_k + t_k d_k) \geq \overline{\eta} \|g_k\|^2 \cos^2 \angle(d_k, -g_k), \tag{35}$$

where the constant $\overline{\eta}$ does not depend on k.

Theorem 1.2.5. [27] *Let $f : \mathbb{R}^n \to \mathbb{R}$ be twice continuously differentiable on open convex set $D \subset \mathbb{R}^n$. Assume that for any $x_0 \in D$ there exists a constant $m > 0$, such that $f(x)$ satisfies the relation (34) on the level set $L(x_0)$. If the line search satisfies the relation (35), then the sequence $\{x_k\}$, generated by Algorithm 1.2.8, with the inexact Wolfe line search, satisfies*

$$\lim_{k \to \infty} \|g_k\| = 0$$

and $\{x_k\}$ converges to the unique minimizer of $f(x)$.

2.4 Modified Newton method

The main problem in Newton method could be the fact that the Hessian G_k may be not positive definite. In that case, we are not sure that the objective function f has its minimizers; furthermore, when G_k is indefinite, the objective function f is unbounded.

So, many modified schemes are made. Now, we describe the next two methods shortly.

In [58], Goldstein and Price use the steepest descent method when G_k is not positive definite. Denoting the angle between d_k and $-g_k$ by θ, as well as having in view the angle rule, $\theta \leq \frac{\pi}{2} - \mu$, where $\mu > 0$, they determine the direction d_k as

$$d_k = \begin{cases} -G_k^{-1} g_k, & \text{if } \cos\theta \geq \eta, \\ -g_k, & \text{otherwise,} \end{cases}$$

where $\eta > 0$ is a given constant.

In [59], the authors present another modified Newton method. When G_k is not positive definite, Hessian G_k is changed into $G_k + \nu_k I$, where $\nu_k > 0$ is chosen in such a way that $G_k + \nu_k I$ is positive definite and well-conditioned. Otherwise, when G_k is positive definite, $\nu_k = 0$.

To consider the other modified Newton methods, such as finite difference Newton method, negative curvature direction method, Gill-Murray stable Newton method, etc., one can see [27], for example.

2.5 Inexact Newton method

By the other side, because of the high cost of the exact Newton method, especially when the dimension n is large, the inexact Newton method might be a good solution. This type of method means that we only approximately solve the Newton equation.

Consider solving the nonlinear equations:

$$F(x) = 0, \tag{36}$$

where $F : \mathbb{R}^n \to \mathbb{R}^n$ is assumed to have the next properties:
A₁ There exists x^* such that $F(x^*) = 0$.
A₂ F is continuously differentiable in the neighborhood of x^*.
A₃ $F'(x^*)$ is nonsingular.
Remind that the basic Newton step is obtained by solving

$$F'(x_k)s_k = -F(x_k)$$

and setting

$$x_{k+1} = x_k + s_k.$$

The inexact Newton method means that we solve

$$F'(x_k)s_k = -F(x_k) + r_k, \tag{37}$$

where

$$\|r_k\| \le \eta_k \|F(x_k)\|. \tag{38}$$

Set

$$x_{k+1} = x_k + s_k. \tag{39}$$

Here, r_k denotes the residual, and the sequence $\{\eta_k\}$, where $0 < \eta_k < 1$, is the sequence which controls the inexactness.

Now, we give two theorems; the first of them claims the linear convergence, and the second claims the superlinear convergence of the inexact Newton method.

Theorem 1.2.6. [27] *Let $F : \mathbb{R}^n \to \mathbb{R}^n$ satisfy the assumptions A_1–A_3. Let the sequence $\{\eta_k\}$ satisfies $0 \le \eta_k \le \eta < t < 1$. Then, for some $\epsilon > 0$, if the starting point is sufficiently near x^*, the sequence $\{x_k\}$ generated by inexact Newton's method (37)–(39) converges to x^*, and the convergence rate is linear, i.e.*

$$\|x_{k+1} - x^*\|_* \le t\|x_k - x^*\|_*,$$

where $\|y\|_* = \|F'(x^*)y\|$.

Theorem 1.2.7. [27] *Let all assumptions of Theorem 1.2.6 hold. Assume that the sequence $\{x_k\}$, generated by the inexact Newton method, converges to x^*. Then*

$$\|r_k\| = o(\|F(x_k)\|), k \to \infty,$$

if and only if $\{x_k\}$ converges to x^* superlinearly.
The relation

$$x_{k+1} = x_k - \frac{f'(x_k)}{f'(x_k) - f'(x_{k-1})} \cdot (x_k - x_{k-1}), \tag{40}$$

presents the secant method.

In [60], a modification of the classical secant method for solving nonlinear, univariate, and unconstrained optimization problems based on the development of the cubic approximation is presented. The iteration formula including an approximation of the third derivative of $f(x)$ by using the Taylor series expansion is derived. The basic assumption on the objective function $f(x)$ is that $f(x)$ is a real-valued function of a single, real variable x and that $f(x)$ has a minimum at x^*. Furthermore, in this chapter it is noted that the secant method is the simplification of Newton method. But, the order of the secant method is lower than one of the Newton methods; it is Q-superlinearly convergent, and its order is $p = \frac{\sqrt{5}+1}{2} \approx 1,618$.

This modified secant method is constructed in [60], having in view, as it is emphasized, that it is possible to construct a cubic function which agrees with $f(x)$ up to the third derivatives. The third derivative of the objective function f is approximated as

$$f^{'''}(x) = \frac{3\left\{\frac{2\left[f'(x_k) - \frac{f(x_k)-f(x_{k-1})}{x_k - x_{k-1}}\right]}{x_k - x_{k-1}} - f^{''}(x_k)\right\}}{x_{k-1} - x_k}.$$

In [61], the authors propose an inexact Newton-like conditional gradient method for solving constrained systems of nonlinear equations. The local convergence of the new method as well as results on its rate is established by using a general majorant condition.

2.6 Quasi-Newton method

Consider the Newton method.

For various practical problems, the computation of Hessian may be very expensive, or difficult, or Hessian can be unavailable analytically. So, the class of so-called quasi-Newton methods is formed, such that it uses only the objective function values and the gradients of the objective function and it is close to Newton method. Quasi-Newton method is such a class of methods which does not compute Hessian, but it generates a sequence of Hessian approximations and maintains a fast rate of convergence.

So, we would like to construct Hessian approximation B_k in quasi-Newton method. Naturally, it is desirable that the sequence $\{B_k\}$ possesses positive definiteness, as well as its direction $d_k = -B_k^{-1}g_k$ should be a descent one.

Now, let $f : \mathbb{R}^n \to \mathbb{R}$ be twice continuously differentiable function on an open set $D \subset \mathbb{R}^n$. Consider the quadratic approximation of f at x_{k+1}:

$$f(x) \approx f(x_{k+1}) + g_{k+1}^T(x - x_{k+1}) + \frac{1}{2}(x - x_{k+1})^T G_{k+1}(x - x_{k+1}).$$

Finding the derivatives, we get

$$g(x) \approx g_{k+1} + G_{k+1}(x - x_{k+1}).$$

Setting $x = x_k$ and using the standard notation: $s_k = x_{k+1} - x_k, y_k = g_{k+1} - g_k$, from the last relation, we get

$$G_{k+1}^{-1} y_k \approx s_k. \tag{41}$$

Relation (41) transforms into the next one if f is the quadratic function:

$$G_{k+1}^{-1} y_k = s_k. \tag{42}$$

Let H_k be the approximation of the inverse of Hessian. Then, we want H_k to satisfy the relation (42). In this way, we come to the quasi-Newton condition or quasi-Newton equation:

$$H_{k+1} y_k = s_k. \tag{43}$$

Let $B_{k+1} = H_{k+1}^{-1}$ be the approximation of Hessian G_{k+1}. Then

$$B_{k+1} s_k = y_k \tag{44}$$

is also the quasi-Newton equation.

If

$$s_k^T y_k > 0, \tag{45}$$

then the matrix B_{k+1} is positive definite. The condition (45) is known as the curvature condition.

Algorithm 1.2.9. *(A general quasi-Newton method).*
Assumptions: $0 \le \epsilon < 1$, $x_0 \in \mathbb{R}^n$, $H_0 \in \mathbb{R}^{n \times n}$. Let $k = 0$.
Step 1. If $\|g_k\| \le \epsilon$, then STOP.
Step 2. Compute $d_k = -H_k g_k$.
Step 3. Find t_k by line search and set $x_{k+1} = x_k + t_k d_k$.
Step 4. Update H_k into H_{k+1} such that quasi-Newton equation (43) holds.
Step 5. Set $k = k + 1$ and go to Step 1.
In Algorithm 1.2.9, usually we take $H_0 = I$, where I is an identity matrix.
Sometimes, instead of H_k, we use B_k in Algorithm 1.2.9.
Then, *Step* 2 becomes
Step 2*. Solve

$$B_k d = -g_k, \text{ for } d_k.$$

By the other side, *Step* 4 becomes
Step 4*. Update B_k into B_{k+1} in such a way that quasi-Newton equation (44) holds.

2.7 Symmetric rank-one ($SR1$) update

Let H_k be the inverse Hessian approximation of the kth iteration. We are trying to update H_k into H_{k+1}, i.e.

$$H_{k+1} = H_k + E_k,$$

where E_k is a matrix with a lower rank. If it is about a rank-one update, we get

$$H_{k+1} = H_k + uv^T, \tag{46}$$

where $u, v \in \mathbb{R}^n$. Using quasi-Newton equation (43), we can get

$$H_{k+1} y_k = \left(H_k + uv^T\right) y_k = s_k,$$

wherefrom

$$(v^T y_k)u = s_k - H_k y_k. \tag{47}$$

Further, from (46) and (47), we have

$$H_{k+1} = H_k + \frac{1}{v^T y_k}(s_k - H_k y_k)v^T.$$

Having in view that the inverse Hessian approximation H_k has to be the symmetric one, we use $v = s_k - H_k y_k$, so we get the symmetric rank-one update (i.e., *SR*1 update):

$$H_{k+1} = H_k + \frac{(s_k - H_k y_k)(s_k - H_k y_k)^T}{(s_k - H_k y_k)^T y_k}. \tag{48}$$

Theorem 1.2.8. [27] *(Property theorem of SR1 update) Let s_0, s_1, and s_{n-1} be linearly independent. Then, for quadratic function with a positive definite Hessian, SR1 method terminates at $n+1$ steps, i.e., $H_n = G^{-1}$.*

More information about SR1 update can be found.

2.8 Davidon-Fletcher-Powell (*DFP*) update

There exists another type of update, which is a rank-two update. In fact, we get H_{k+1} using two symmetric, rank-one matrices:

$$H_{k+1} = H_k + auu^T + bvv^T, \tag{49}$$

where $u, v \in \mathbb{R}^n$ and a, b are scalars which have to be determined.

Using quasi-Newton equation (43), we can get

$$H_k y_k + auu^T y_k + bvv^T y_k = s_k. \tag{50}$$

The values of u, v are not determined in a unique way, but the good choice is

$$u = s_k, v = H_k y_k.$$

Now, from (50), we get:

$$a = \frac{1}{s_k^T y_k}, b = -\frac{1}{y_k^T H_k y_k}.$$

Hence, we get the formula

$$H_{k+1} = H_k + \frac{s_k s_k^T}{s_k^T y_k} - \frac{H_k y_k y_k^T H_k}{y_k^T H_k y_k}, \tag{51}$$

which is *DFP* update.

Theorem 1.2.9. [27] *(Positive definiteness of DFP update) DFP update (51) retains positive definiteness if and only if $s_k^T y_k > 0$.*

Theorem 1.2.10. [27] *(Quadratic termination theorem of DFP method) Let $f(x)$ be a quadratic function with positive definite Hessian G. Then, if the exact line search is used, the sequence $\{s_j\}$, generated from DFP method, satisfies, for $i = 0, 1, \ldots, m$, where $m \leq n-1$:*

1. $H_{i+1}y_j = s_j, j = 0, 1, \ldots, i$ (*hereditary property*).

2. $s_i^T G s_j = 0, j = 0, 1, \ldots, i-1$ (*conjugate direction property*).

3. *The method terminates at* $m + 1 \leq n$ *steps. If* $m = n - 1$, *then* $H_n = G^{-1}$.

2.9 Broyden-Fletcher-Goldfarb-Shanno (*BFGS*) update

BFGS update is given by the formula

$$B_{k+1}^{BFGS} = B_k + \frac{y_k y_k^T}{y_k^T s_k} - \frac{B_k s_k s_k^T B_k}{s_k^T B_k s_k}. \tag{52}$$

The *BFGS* update is also said to be a complement to *DFP* update.

In [62], an adaptive scaled *BFGS* method for unconstrained optimization is presented. In this paper, the author emphasizes that the *BFGS* method is one of the most efficient quasi-Newton methods for solving small-size and medium-size unconstrained optimization problems. The third term in the standard *BFGS* update formula is scaled in order to reduce the large eigenvalues of the approximation to the Hessian of the minimizing function. In fact, in [62], the general scaling *BFGS* updating formula is considered:

$$B_{k+1} = B_k - \frac{B_k s_k s_k^T B_k}{s_k^T B_k s_k} + \gamma_k \frac{y_k y_k^T}{y_k^T s_k}, \tag{53}$$

where γ_k is a positive parameter. Obviously, using $\gamma_k = 1$ for all $k = 0, 1, \ldots$, we get the standard *BFGS* formula. By the way, there exist several procedures created to select the scaling parameter γ_k, for example, see [62–69]. The approach for determining the scaling parameters of the terms of the *BFGS* update in [62] is to minimize the Byrd and Nocedal measure function.

Namely, in [70], the next function was introduced:

$$\varphi(A) = tr(A) - \ln(\det(A)), \tag{54}$$

which is defined on positive definite matrices.

This function is a measure of matrices involving all the eigenvalues of A, not only the smallest one and the largest one, as it is traditionally used in the analysis of the quasi-Newton method based on the condition number of matrices.

Observe that function φ works simultaneously with the trace and the determinant, thus simplifying the analysis of the quasi-Newton methods. Fletcher [71] proves that this function is strictly convex on the set of symmetric and positive definite matrices, and it is minimized by $A = I$. Besides, this function becomes unbounded when A becomes singular or infinite, and therefore it works as a barrier function that keeps A positive definite. It is worth saying that the *BFGS* update tends to generate updates with large eigenvalues.

Further, in [62], a double-parameter scaling *BFGS* update is considered, in which the first two terms on the right-hand side of the *BFGS* update (52) are scaled with a positive parameter, while the third one is scaled with another positive parameter:

$$B_{k+1} = \delta_k \left[B_k - \frac{B_k s_k s_k^T B_k}{s_k^T B_k s_k} \right] + \gamma_k \frac{y_k y_k^T}{y_k^T s_k}, \tag{55}$$

where δ_k and γ_k are the two positive parameters that have to be determined.
In [62], the next proposition is proved.

Proposition 1.2.1. *If the step size t_k is determined by the standard Wolfe line search (12) and (13), B_k is positive definite and $\gamma_k > 0$, and then B_{k+1}, given by (55), is also positive definite.*

From (55), it can be seen that $\varphi(B_{k+1})$ depends on the scaling parameters δ_k and γ_k. In [62], these scaling parameters are determined as solution of the minimizing problem:

$$\min_{\delta_k > 0,\, \gamma_k > 0} \varphi(B_{k+1}). \tag{56}$$

Further, the next values of the scaling parameters δ_k and γ_k are reached:

$$\delta_k = \frac{n-1}{tr(B_k) - \frac{\|B_k s_k\|^2}{s_k^T B_k s_k}} \tag{57}$$

$$\gamma_k = \frac{y_k^T s_k}{\|y_k\|^2}. \tag{58}$$

Consider the relation

$$x_{k+1} = x_k + t_k d_k, \tag{59}$$

where d_k is the *BFGS* search direction obtained as solution of the linear algebraic system

$$B_k d_k = -g_k,$$

where the matrix B_k is the *BFGS* approximation to the Hessian $\nabla^2 f(x_k)$, being updated by the classical formula (52).

The next theorems are also given in [62].

Theorem 1.2.11. *If the step size in (59) is determined by the Wolfe search conditions (12)–(13), then the scaling parameters given by (57) and (58) are the unique global solutions of the problem (56).*

Theorem 1.2.12. *Let δ_k be computed by (57). Then, for any $k = 0, 1, \ldots$, δ_k is positive and close to 1.*

Next, in [72], using chain rule, a modified secant equation is given, to get a more accurate approximation of the second curvature of the objective function. Then, based on this modified secant equation, a new *BFGS* method is presented. The proposed method makes use of both gradient and function values, and it utilizes information from two most recent steps, while the usual secant relation uses only the latest step information. Under appropriate conditions, it is shown that the proposed method is globally convergent without convexity assumption on the objective function.

Some interesting applications of Newton, modified Newton, inexact Newton, and quasi-Newton methods can be found, for example, in [73–83], etc.

A very interesting paper is [84].

An interesting application of *BFGS* method can be found in [85].

3. Conclusion

Today, the modifications of the line search techniques are very actual and all in the aim to create new, better optimization methods.

Further, following recent trends in unconstrained optimization, we can notice that almost all optimization methods, which are considered in this chapter, are still actual.

They are applied in the other areas of Mathematics, as well as in practice. Also, different modifications of these methods are made, in the aim to improve them.

Let us emphasize that *BFGS* update is very popular now.

Author details

Snezana S. Djordjevic
Faculty of Technology, University of Nis, Leskovac, Serbia

*Address all correspondence to: snezanadjordjevicle@gmail.com

References

[1] Armijo L. Minimization of functions having Lipschitz continuous first partial derivatives. Pacific Journal of Mathematics. 1966;**16**(1):1-3

[2] Goldstein AA. On steepest descent. SIAM Journal on Control and Optimization. 1965;**3**:147-151

[3] Wolfe P. Convergence conditions for ascent methods. SIAM Review. 1969;**11**: 226-235

[4] Wolfe P. Convergence conditions for ascent methods. II: Some corrections. SIAM Review. 1969;**11**:226-235

[5] Nocedal J, Wright SJ. Numerical Optimization. New York, NY, USA: Springer Verlag; 2006

[6] Krejic N, Jerinkic NK. Nonmonotone line search methods with variable sample size. Numerical Algorithms. 2015;**68**(4):711-739

[7] Dai YH. On the nonmonotone line search. Journal of Optimization Theory and Applications. 2002;**112**:315-330

[8] Raydan M. The Barzilai and Borwein gradient method for the large scale unconstrained minimization problem. SIAM Journal on Optimization. 1997;**7**: 26-33

[9] Toint PhL. Nonmonotone trust region algorithms for nonlinear optimization subject to convex constraints. Mathematical Programming. 1997;**77**:69. DOI: 10.1007/BF02614518

[10] Toint PL. An assessment of non-monotone line search techniques for unconstrained optimization. SIAM Journal on Scientific Computing. **17**(3): 725-739. 15 pages

[11] Zhang H, Hager W. A nonmonotone line search technique and its application to unconstrained optimization. SIAM Journal on Optimization. 2004;**4**: 1043-1056

[12] Grippo L, Lampariello F, Lucidi S. A nonmonotone line search technique for Newton's method. SIAM Journal on Numerical Analysis. 1986;**23**:707-716

[13] Li DH, Fukushima M. A derivative-free line search and global convergence of Broyden-like method for nonlinear equations. Optimization Methods and Software. 2000;**13**:181-201

[14] Birgin EG, Krejic N, Martinez JM. Globally convergent inexact Quasi-Newton methods for solving nonlinear systems. Numerical Algorithms. 2003; **32**:249-250

[15] SHI Z-J, Shen J. Convergence of descent method with new line search. Journal of Applied Mathematics and Computing. 2006;**20**(1–2):239-254

[16] Wan et al. New cautious BFGS algorithm based on modified Armijo-type line search. Journal of Inequalities and Applications. 2012;**2012**:241

[17] Yu G, Guan L, Wei Z. Globally convergent Polak-Ribiére-Polyak conjugate gradient methods under a modified Wolfe line search. Applied Mathematics and Computation. 2009; **215**:3082-3090

[18] Huang S, Wan Z, Zhang J. An extended nonmonotone line search technique for large-scale unconstrained optimization. Journal of Computational and Applied Mathematics. 2018;**330**: 586. 19p

[19] Koorapetse MS, Kaelo P. Globally convergent three-term conjugate

gradient projection methods for solving nonlinear monotone equations. Arabian Journal of Mathematics. 2018;**7**(4): 289-301

[20] Yu Z, Pu D. A new nonmonotone line search technique for unconstrained optimization. Journal of Computational and Applied Mathematics. 2008;**219**: 134-144

[21] Yuan G, Wei Z. A modified PRP conjugate gradient algorithm with nonmonotone line search for nonsmooth convex optimization problems. Journal of Applied Mathematics and Computing. 2016;**51**: 397-412

[22] Yuan G, Wei Z, Lu X. Global convergence of the BFGS method and the PRP method for general functions under a modified weak Wolfe-Powell line search. Applied Mathematical Modelling. 2017;**47**: 811-825

[23] Yuan G, Wei Z, Yang Y. The global convergence of the Polak-Ribiére-Polyak conjugate gradient algorithm under inexact line search for nonconvex functions. Journal of Computational and Applied Mathematics. 2018. DOI: 10.1016/j.cam.2018.10.057. In press

[24] Cauchy A. Méthode générale pour la résolution des systéms d'equations simultanées. Comptes Rendus Mathematique Academie des Sciences, Paris. 1847;**25**:46-89

[25] Johari R, Rivaie M, Mamat M. A new scaled steepest descent method for unconstrained optimization with global convergence properties. Journal of Engineering and Applied Sciences. 2018; **13**(Special Issue 6):5442-5445

[26] Wen GK, Mamat M, Mohd IB, Dasril Y. A novel of step size selection procedures for steepest descent method. Applied Mathematical Sciences. 2012; **6**(51):2507-2518

[27] Sun W, Yuan Y-X. Optimization theory and methods: Nonlinear programming. Springer: Optimization and Its Applications. 2006

[28] Abidin ZAZ, Mamat M, Rivaie M, Mohd I. A new steepest descent method. In: Proceedings of the 3rd International Conference on Mathematical Sciences, Vol. 1602, December 17–19; Melville, New York: AIP; 2013. pp. 273-278

[29] Andrei N. Relaxed Gradient Descent and a New Gradient Descent Methods for Unconstrained Optimization. Available from: https://camo.ici.ro/neculai/newgrad.pdf

[30] Knyazev AV, Lashuk I. Steepest descent and conjugate gradient methods with variable preconditioning. SIAM Journal on Matrix Analysis and Applications. 2007;**29**(4):1267-1280

[31] Liu C-S, Chang J-R, Chen Y-W. A modified algorithm of steepest descent method for solving unconstrained nonlinear optimization problems. Journal of Marine Science and Technology. 2015;**23**(1):88-97

[32] Osadcha O, Marszaek Z. Comparison of Steepest Descent Method and Conjugate Gradient Method. Available from: http://ceur-ws.org/Vol-1853/p01.pdf

[33] Andrei N. Scaled conjugate gradient algorithms for unconstrained optimization. Computational Optimization and Applications. 2007;**38**: 401-416

[34] Napitupulu et al. Steepest descent method implementation on unconstrained optimization problem using C++ program. IOP Conference

Series: Materials Science and Engineering. 2018;**332**:012024

[35] Yuan Y. A new stepsize for the steepest descent method. Journal of Computational Mathematics. 2006; **24**(2):149-156

[36] Raydan M, Svaiter B. Relaxed steepest descent and Cauchy-Barzilai-Borwein method. Computational Optimization and Applications. 2002; **21**(2):155-167

[37] Cai Y, Bai Z, Pask JE, Sukumar N. Convergence analysis of a locally accelerated preconditioned steepest descent method for hermitian-definite generalized eigenvalue problems. Journal of Computational Mathematics. 2018;**36**(5):739-760

[38] Egorova I, Michor J, Teschl G. Rarefaction waves for the Toda equation via nonlinear steepest descent. Discrete and Continuous Dynamical Systems. 2018;**38**:2007-2028

[39] Gonzaga CC. On the worst case performance of the steepest descent algorithm for quadratic functions. Mathematical Programming, Series A. 2016;**160**:307-320

[40] Hosokawa S, Pusztai L, Matsushita T. Algorithm for atomic resolution holography using modified $L1$-regularized linear regression and steepest descent method. Physica Status Solidi B: Basic Solid State Physics. 2018; **255**:11

[41] Liu X, Reynolds AC. A multiobjective steepest descent method with applications to optimal well control. Computational Geosciences. 2016;**20**:355-374

[42] Svaiter BF. Hölder continuity of the steepest descent direction for multiobjective optimization. 2018. arXiv:1802.01402v1 [math.OC]

[43] Torres P, van Wingerden J-W. Identification of 2D interconnected systems: An efficient steepest-descent approach. IFAC Papers OnLine. 2018; **51**(15):78-83

[44] Barzilai J, Borwein J. Two-point step size gradient methods. IMA Journal of Numerical Analysis. 1988;**8**(1):141-148

[45] Grippo L, Sciandrone M. Nonmonotone globalization techniques for the Barzilai-Borwein gradient method. Computational Optimization and Applications. 2002;**23**:143-169

[46] Dai YH, Hager WW, Schittkowski K, Zhang H. The cyclic Barzilai-Borwein method for unconstrained optimization. IMA Journal of Numerical Analysis. 2006;**26**:604-627

[47] Raydan M. On the Barzilai and Borwein choice of steplength for the gradient method. IMA Journal of Numerical Analysis. 1993;**13**(3):321-326

[48] Dai Y, Liao L. R-linear convergence of the Barzilai and Borwein gradient method. IMA Journal of Numerical Analysis. 2002;**22**(1):1-10

[49] Friedlander A, Martinez JM, Molina B, Raydan M. Gradient method with retards and generalizations. SIAM Journal on Numerical Analysis. 1999;**36**: 275-289

[50] Narushima Y, Wakamatsu T, Yabe H. Extended Barzilai-Borwein method for unconstrained minimization problems. Pacific Journal of Optimization. 2008;**6**(3):591-614

[51] Dai YH, Yuan Y. Analysis of monotone gradient methods. Journal of Industrial and Management Optimization. 2005;**1**:181-192

[52] Fletcher R. On the Barzilai-Borwein method, Optimization and Control with Applications. Springer Series in Applied Optimization 96. New York: Springer-Verlag; 2005. pp. 235-256

[53] Liu ZX, Liu HW. An efficient gradient method with approximate optimal stepsize for large-scale unconstrained optimization. Numerical Algorithms. 2018;**78**(1):21-39

[54] La Cruz W. A spectral algorithm for large-scale systems of nonlinear monotone equations. Numerical Algorithms. 2017;**76**:1109-1130

[55] Feng X, Hormuth DA II, Yankeelov TE. An adjoint-based method for a linear mechanically-coupled tumor model: Application to estimate the spatial variation of murine glioma growth based on diffusion weighted magnetic resonance imaging. Computational Mechanics. 2018. DOI: 10.1007/s00466-018-1589-2

[56] Krzysztof S, Drozda Stochastic P. Gradient descent with Barzilai-Borwein update step for *SVM*. Information Sciences. 2015;**316**:218-233

[57] Li M, Liu H, Liu Z. A new subspace minimization conjugate gradient method with nonmonotone line search for unconstrained optimization. Numerical Algorithms. 2018;**79**:195-219

[58] Goldstein AA, Price JF. An effective algorithm for minimization. Numerische Mathematik. 1967;**10**:184-189

[59] Goldfeld SM, Quandt RE, Trotter HF. Maximisation by quadratic hill-climbing. Econometrica. 1966;**34**: 541-551

[60] Kahya E, Chen J. A modified Secant method for unconstrained optimization. Applied Mathematics and Computation. 2007;**186**:1000-1004

[61] Gonçalves MLN, Oliveira FR. An inexact Newton-like conditional gradient method for constrained nonlinear systems. Applied Numerical Mathematics. 2018;**132**:22-34

[62] Andrei N. An adaptive scaled BFGS method for unconstrained optimization. Numerical Algorithms. 2018;**77**(2): 413-432

[63] Andrei N. A double parameter scaled BFGS method for unconstrained optimization. Journal of Computational and Applied Mathematics. 2018;**332**:26-44

[64] Biggs MC. Minimization algorithms making use of non-quadratic properties of the objective function. Journal of the Institute of Mathematics and its Applications. 1971;**8**:315-327

[65] Biggs MC. A note on minimization algorithms making use of non-quadratic properties of the objective function. Journal of the Institute of Mathematics and its Applications. 1973; **12**:337-338

[66] Liao A. Modifying *BFGS* method. Operations Research Letters. 1997;**20**: 171-177

[67] Nocedal J, Yuan YX. Analysis of self-scaling quasi-Newton method. Mathematical Programming. 1993;**61**: 19-37

[68] Oren SS, Luenberger DG. Self-scaling variable metric (SSVM) algorithms, Part I: Criteria and sufficient conditions for scaling a class of algorithms. Management Science. 1974; **20**:845-862

[69] Yuan YX. A modified BFGS algorithm for unconstrained optimization. IMA Journal of Numerical Analysis. 1991;**11**:325-332

[70] Byrd R, Nocedal J. A tool for the analysis of quasi-Newton methods with application to unconstrained minimization. SIAM Journal on Numerical Analysis. 1989;**26**:727-739

[71] Fletcher R. An overview of unconstrained optimization. In: Spedicato E, editor. Algorithms for Continuous Optimization: The State of the Art. Boston: Kluwer Academic Publishers; 1994. pp. 109-143

[72] Dehghani R, Bidabadi N, Hosseini MM. A new modified BFGS method for unconstrained optimization problems. Computational and Applied Mathematics. 2018;**37**:5113-5125

[73] Andrei N. A diagonal quasi-Newton updating method for unconstrained optimization. Numerical Algorithms. 2018:16. DOI: 10.1007/s11075-018-0562-7. In press

[74] Bajović D, Jakovetić D, Krejić N, Krklec Jerinkić N. Newton-like method with diagonal correction for distributed optimization. SIAM Journal on Optimization. 2017;**27**(2):1171-1203

[75] Carraro T, Dörsam S, Frei S, Schwarz D. An adaptive newton algorithm for optimal control problems with application to optimal electrode design. Journal of Optimization Theory and Applications. 2018;**177**:498-534

[76] Djordjević SS. Two modifications of the method of the multiplicative parameters in descent gradient methods. Applied Mathematics and Computation. 2012;**218**(17):8672-8683

[77] Ferreira OP, Silva GN. Inexact Newton method for non-linear functions with values in a cone. Applicable Analysis. 2018. https://www.tandfonline.com/doi/abs/10.1080/00036811.2018.1430779

[78] Grapsa TN. A modified Newton direction for unconstrained optimization. A Journal of Mathematical Programming and Operations Research. 2014;**63**(7):983-1004

[79] Li Y-M, Guo X-P. On the accelerated modified Newton-HSS method for systems of nonlinear equations. Numerical Algorithms. 2018;**79**:1049-1073

[80] Matebese B, Withey D, Banda MK. Modified Newton's method in the leapfrog method for mobile robot path planning. In: Dash S, Naidu P, Bayindir R, Das S, editors. Artificial Intelligence and Evolutionary Computations in Engineering Systems. Advances in Intelligent Systems and Computing. Vol. 668. Singapore: Springer; 2018. pp. 71-78

[81] Mezzadri F, Galligani E. An inexact Newton method for solving complementarity problems in hydrodynamic lubrication. Calcolo. 2018;**55**:1

[82] Sharma JR, Argyros IK, Kumar D. Newton-like methods with increasing order of convergence and their convergence analysis in Banach space. SeMA. 2018;**75**:545-561

[83] Stanimirović P, Miladinović M, Djordjević S. Multiplicative parameters in gradient descent methods. Univerzitet u Nišu. 2009;**23**(3):23-36

[84] Petrović MJ, Stanimirović PS, Kontrec N, Mladenov J. Hybrid modification of accelerated double direction method. Mathematical Problems in Engineering. 2018;**2018**:1-8

[85] Stanimirovic PS, Ivanov B, Djordjevic S, Brajevic I. New hybrid conjugate gradient and Broyden-Fletcher-Goldfarb-Shanno conjugate gradient methods. Journal of Optimization Theory and Applications. 2018;**178**(3):860-884

Chapter 2

Unconstrained Optimization Methods: Conjugate Gradient Methods and Trust-Region Methods

Snezana S. Djordjevic

Abstract

Here, we consider two important classes of unconstrained optimization methods: conjugate gradient methods and trust region methods. These two classes of methods are very interesting; it seems that they are never out of date. First, we consider conjugate gradient methods. We also illustrate the practical behavior of some conjugate gradient methods. Then, we study trust region methods. Considering these two classes of methods, we analyze some recent results.

Keywords: conjugate gradient method, hybrid conjugate gradient method, three-term conjugate gradient method, modified conjugate gradient method, trust region methods

1. Introduction

Remind to the unconstrained optimization problem which we can present as

$$\min_{x \in R^n} f(x), \tag{1}$$

where $f : \mathbb{R}^n \to \mathbb{R}$ is a smooth function.

Here, we consider two classes of unconstrained optimization methods: conjugate gradient methods and trust region methods. Both of them are made with the aim to solve the unconstrained optimization problem (1).

In this chapter, at first, we consider the conjugate gradient methods. Then, we study trust region methods. Also, we try to give some of the most recent results in these areas.

2. Conjugate gradient method (shortly *CG*)

The conjugate gradient method is the method between the steepest descent method and the Newton method.

The conjugate gradient method in fact deflects the direction of the steepest descent method by adding to it a positive multiple of the direction used in the last step.

The restarting and the preconditioning are very important to improve the conjugate gradient method [47].

Some of well-known *CG* methods are [12, 19, 20, 23, 24, 31, 39, 40, 49]:

$$\beta_k^{HS} = \frac{g_{k+1}^T y_k}{d_k^T y_k}$$

$$\beta_k^{FR} = \frac{\|g_{k+1}\|^2}{\|g_k\|^2}$$

$$\beta_k^{PRP} = \frac{g_{k+1}^T y_k}{\|g_k\|^2}$$

$$\beta_k^{CD} = \frac{\|g_{k+1}\|^2}{-d_k^T g_k}$$

$$\beta_k^{LS} = \frac{g_{k+1}^T y_k}{-d_k^T g_k}$$

$$\beta_k^{DY} = \frac{\|g_{k+1}\|^2}{d_k^T y_k}$$

$$\beta_k^{N} = \left(y_k - 2d_k \frac{\|y_k\|^2}{d_k^T y_k}\right)^T \frac{g_{k+1}}{d_k^T y_k}$$

$$\beta_k^{WYL} = \frac{g_k^T \left(g_k - \frac{\|g_k\|}{\|g_{k-1}\|} g_{k-1}\right)}{\|g_{k-1}\|^2}$$

Consider positive definite quadratic function

$$f(x) = \frac{1}{2} x^T G x + b^T x + c, \tag{2}$$

where G is an $n \times n$ symmetric positive definite matrix, $b \in \mathbb{R}^n$, and c is a real number.

Theorem 1.2.1. *[47] (Property theorem of conjugate gradient method) For positive definite quadratic function (2), FR conjugate gradient method with the exact line search terminates after $m \leq n$ steps, and the following properties hold for all i, $0 \leq i \leq m$:*

$$d_i^T G d_j = 0, j = 0, 1, ..., i-1;$$

$$g_i^T g_j = 0, j = 0, 1, ..., i-1;$$

$$d_i^T g_i = -g_i^T g_i;$$

$$[g_0, g_1, ..., g_i] = [g_0, Gg_0, ..., G^i g_0];$$

$$[d_0, d_1, ..., d_i] = [g_0, Gg_0, ..., G^i g_0],$$

where m is the number of distinct eigenvalues of G.

Now, we give the algorithm of conjugate gradient method.

Algorithm 1.2.1. *(CG method).*

Assumptions: $\varepsilon < 0$ and $x_0 \in \mathbb{R}^n$. *Let* $k = 0$, $t_0 = 0$, $d_{-1} = 0$, $d_0 = -g_0$, $\beta_{-1} = 0$, and $\beta_0 = 0$.

Step 1. If $\|g_k\| \leq \varepsilon$*, then STOP.*
Step 2. Calculate the step-size t_k *by a line search.*
Step 3. Calculate β_k *by any of the conjugate gradient method.*
Step 4. Calculate $d_k = -g_k + \beta_{k-1} d_{k-1}$.
Step 5. Set $x_{k+1} = x_k + t_k d_k$.
Step 6. Set $k = k + 1$ *and go to Step 1.*

2.1 Convergence of conjugate gradient methods

Theorem 1.2.2. *[47] (Global convergence of FR conjugate gradient method) Suppose that* $f : \mathbb{R}^n \rightarrow \mathbb{R}$ *is continuously differentiable on a bounded level set*

$$L = \{x \in \mathbb{R}^n | f(x) \leq f(x_0)\},$$

and let FR method be implemented by the exact line search. Then, the produced sequence $\{x_k\}$ has at least one accumulation point, which is a stationary point, i.e.:

1. *When* $\{x_k\}$ *is a finite sequence, then the final point* x^* *is a stationary point of* f.

2. *When* $\{x_k\}$ *is an infinite sequence, then it has a limit point, and it is a stationary point.*

In [35], a comparison of two methods, the steepest descent method and the conjugate gradient method which are used for solving systems of linear equations, is illustrated. The aim of the research is to analyze, which method is faster in solving these equations and how many iterations are needed by each method for solving.

The system of linear equations in the general form is considered:

$$Ax = B, \tag{3}$$

where matrix A is symmetric and positive definite.

The conclusion is that the *SD* method is a faster method than the *CG*, because it solves equations in less amount of time.

By the other side, the authors find that the *CG* method is slower but more productive than the *SD*, because it converges after less iterations.

So, we can see that one method can be used when we want to find solution very fast and another can converge to maximum in less number of iterations.

Again, we consider the problem (1), where $f : \mathbb{R}^n \rightarrow \mathbb{R}$ is a smooth function and its gradient is available.

A hybrid conjugate gradient method is a certain combination of different conjugate gradient methods; it is made to improve the behavior of these methods and to avoid the jamming phenomenon.

An excellent survey of hybrid conjugate gradient methods is given in [5].

Three-term conjugate gradient methods were studied in the past (e.g., see [8, 32, 34], etc.); but, from recent papers about *CG* methods, we can conclude that maybe the mainstream is made by three-term and even four-term conjugate gradient methods. An interesting paper about a five-term hybrid conjugate gradient method is [1]. Also, from recent papers we can conclude that different modifications of the existing *CG* methods are made, as well as different hybridizations of *CG* and *BFGS* methods.

Consider unconstrained optimization problem (1), where $f : \mathbb{R}^n \rightarrow \mathbb{R}$ is a continuously differentiable function, bounded from below. Starting from an initial point $x_0 \in \mathbb{R}^n$, the three-term conjugate gradient method with line search generates a sequence $\{x_k\}$, given by the next iterative scheme:

$$x_{k+1} = x_k + t_k d_k, \quad (4)$$

where t_k is a step-size which is obtained from the line search, and

$$d_0 = -g_0, d_{k+1} = -g_{k+1} + \delta_k s_k + \eta_k y_k.$$

In the last relation, δ_k and η_k are the conjugate gradient parameters, $s_k = x_{k+1} - x_k, g_k = \nabla f(x_k)$, and $y_k = g_{k+1} - g_k$. We can see that the search direction d_{k+1} is computed as a linear combination of $-g_{k+1}$, s_k, and y_k.

In [6], the author suggests another way to get three-term conjugate gradient algorithms by minimization of the one-parameter quadratic model of the function f. The idea is to consider the quadratic approximation of the function f in the current point and to determine the search direction by minimization of this quadratic model. It is assumed that the symmetrical approximation of the Hessian matrix B_{k+1} satisfies the general quasi-Newton equation which depends on a positive parameter:

$$B_{k+1} s_k = \omega^{-1} y_k, \omega \neq 0. \quad (5)$$

In this paper the quadratic approximation of the function f is considered:

$$\Phi_{k+1}(d) = f_{k+1} + g_{k+1}^T d + \frac{1}{2} d^T B_{k+1} d.$$

The direction d_{k+1} is computed as

$$d_{k+1} = -g_{k+1} + \beta_k s_k, \quad (6)$$

where the scalar β_k is determined as the solution of the following minimizing problem:

$$\min_{\beta_k \in \mathbb{R}} \Phi_{k+1}(d_{k+1}). \quad (7)$$

From (6) and (7), the author obtains

$$\beta_k = \frac{g_{k+1}^T B_{k+1} s_k - g_{k+1}^T s_k}{s_k^T B_{k+1} s_k}. \quad (8)$$

Using (5), from (7), the next expression for β_k is obtained:

$$\beta_k = \frac{g_{k+1}^T y_k - \omega g_{k+1}^T s_k}{y_k^T s_k}. \quad (9)$$

Using the idea of Perry [36], the author obtains

$$d_{k+1} = -g_{k+1} + \frac{y_k^T g_{k+1} - \omega s_k^T g_{k+1}}{y_k^T s_k} s_k - \frac{s_k^T g_{k+1}}{y_k^T s_k} y_k.$$

In fact, in this approach the author gets a family of three-term conjugate gradient algorithms depending of a positive parameter ω.

Next, in [52], the *WYL* conjugate gradient (*CG*) formula, with $\beta_k^{WYL} \geq 0$, is further studied. A three-term *WYL CG* algorithm is presented, which has the sufficiently descent property without any conditions. The global convergence and the linear convergence are proven; moreover, the n-step quadratic convergence with a restart strategy is established if the initial step length is appropriately chosen.

The first three-term Hestenes-Stiefel (*HS*) method (*TTHS* method) can be found in [55].

Baluch et al. [7] describe a modified three-term Hestenes-Stiefel (*HS*) method. Although the earliest conjugate gradient method *HS* achieves global convergence using an exact line search, this is not guaranteed in the case of an inexact line search. In addition, the *HS* method does not usually satisfy the descent property. The modified three-term conjugate gradient method from [7] possesses a sufficient descent property regardless of the type of line search and guarantees global convergence using the inexact Wolfe-Powell line search [50, 51]. The authors also prove the global convergence of this method. The search direction, which is considered in [7], has the next form:

$$d_k = \begin{cases} -g_k, \text{ if } k = 0, \\ -g_k + \beta_k^{BZA} d_{k-1} - \theta_k^{BZA} y_{k-1}, \text{ if } k \geq 1, \end{cases}$$

where $\beta_k^{BZA} = \frac{g_k^T(g_k - g_{k-1})}{d_{k-1}^T y_{k-1} + \mu |g_k^T d_{k-1}|}$, $\theta_k^{BZA} = \frac{g_k^T d_{k-1}}{d_{k-1}^T y_{k-1} + \mu |g_k^T d_{k-1}|}$, $\mu > 1$.

In [13], an accelerated three-term conjugate gradient method is proposed, in which the search direction satisfies the sufficient descent condition as well as extended Dai-Liao conjugacy condition:

$$d_k^T y_{k-1} = -t g_k^T s_{k-1}, \ t \geq 0.$$

This method seems different from the existent methods.

Next, Li-Fushikuma quasi-Newton equation is

$$\nabla^2 f(x_k) s_{k-1} = z_{k-1}, \tag{10}$$

where

$$z_{k-1} = y_{k-1} + C \|g_{k-1}\|^r s_{k-1} + \max\left\{ -\frac{s_{k-1}^T y_{k-1}}{\|s_{k-1}\|^2}, 0 \right\} s_{k-1},$$

where C and r are two given positive constants. Based on (10), Zhou and Zhang [56] propose a modified version of *DL* method, called *ZZ* method in [13].

In [30], some new conjugate gradient methods are extended, and then some three-term conjugate gradient methods are constructed. Namely, the authors remind to [41, 42], with its conjugate gradient parameters, respectively:

$$\beta_k^{RMIL} = \frac{g_k^T y_{k-1}}{\|d_{k-1}\|^2}, \tag{11}$$

$$\beta_k^{MRMIL} = \frac{g_k^T (g_k - g_{k-1} - d_{k-1})}{\|d_{k-1}\|^2}, \tag{12}$$

wherefrom it is obvious that $\beta_k^{MRMIL} = \beta_k^{RMIL}$ for the exact line search. Let us say that these methods, presented in [41, 42], are *RMIL* and *MRMIL* methods.

The three-term *RMIL* and *MRMIL* methods are introduced in [30].

The search direction d_k can be expressed as

$$d_0 = -g_0, \ d_k = -g_k + \beta_k d_{k-1} + \theta_k y_{k-1},$$

where β_k is given by (11) or (12), and

$$\theta_k = -\frac{g_k^T d_{k-1}}{\|d_{k-1}\|^2}.$$

An important property of the proposed methods is that the search direction always satisfies the sufficient descent condition without any line search, that is, the next relation always holds

$$g_k^T d_k \leq -\|g_k\|^2.$$

Under the standard Wolfe line search and the classical assumptions, the global convergence properties of the proposed methods are proven.

Having in view the conjugate gradient parameter suggested in [49], in [45] the next two conjugate gradient parameters are presented:

$$\beta_k^{MHS} = \frac{\|g_k\|^2 - \frac{\|g_k\|}{\|g_{k-1}\|} g_k^T g_{k-1}}{d_{k-1}^T (g_k - g_{k-1})}, \tag{13}$$

$$\beta_k^{MLS} = \frac{\|g_k\|^2 - \frac{\|g_k\|}{\|g_{k-1}\|} g_k^T g_{k-1}}{-d_{k-1}^T g_{k-1}}. \tag{14}$$

Motivated by [49], as well as by [45], in [1], a new hybrid nonlinear *CG* method is proposed; it combines the features of five different *CG* methods, with the aim of combining the positive features of different non-hybrid methods. The proposed method generates descent directions independently of the line search. Under some assumptions on the objective function, the global convergence is proven under the standard Wolfe line search. Conjugate gradient parameter, proposed in [1], is

$$\beta_k^{hAO} = \frac{\|g_k\|^2 - \max\left\{0, \frac{\|g_k\|}{\|g_{k-1}\|} g_k^T g_{k-1}\right\}}{\max\left\{\|g_{k-1}\|^2, d_{k-1}^T (g_k - g_{k-1}), -d_{k-1}^T g_{k-1}\right\}}. \tag{15}$$

Let's note that the proposed method is hybrid of *FR*, *DY*, *WYL*, *MHS*, and *MLS*.

The behaviors of the methods *BZA*, *TTRMIL*, *MRMIL*, *MHS*, *MLS*, and *hAO* are illustrated by the next tables.

The test criterion is CPU time.

The tests are performed on the computer Workstation Intel Celeron CPU 1,9 GHz.

The experiments are made on the test functions from [3].

Each problem is tested for a number of variables $n = 1000$ and $n = 5000$.

The average CPU time values are given in the last rows of these tables (**Tables 1–4**).

In [2], based on the numerical efficiency of Hestenes-Stiefel (*HS*) method, a new modified *HS* algorithm is proposed for unconstrained optimization. The new direction independent of the line search satisfies the sufficient descent condition. Motivated by theoretical and numerical features of three-term conjugate gradient (*CG*) methods proposed by [33], similar to the approach in [10], the new direction is computed by minimizing the distance between the *CG* direction and the direction of the three-term *CG* methods proposed by [33]. Under some mild conditions, the global convergence of the new method for general functions is established when the standard Wolfe line search is used. In this paper the conjugate gradient parameter is given by

$$\beta_k = \beta_k^{HS} \theta_k, \tag{16}$$

function	*BZA*	TTRMIL	MRMIL	MHS	MLS	hAO
Ext.Pen.	21.793340	20.966534	16.036903	19.812127	21.933741	20.326930
Pert.Quad.	21.855740	22.542144	15.506499	20.904134	22.230142	18.954121
Raydan1	6.801644	7.066845	6.349241	7.098045	7.066845	7.332047
Raydan2	0.608404	0.592804	0.577204	0.592804	0.608404	0.639604
Diag.1	0.608404	0.608404	0.577204	0.608404	0.514803	0.577204
Diag.2	5.163633	5.600436	4.695630	4.758031	5.662836	4.851631
Diag.3	5.616036	5.694037	5.241634	5.756437	5.584836	5.506835
Gen.Tridiag.-1	3.042019	2.932819	2.683217	2.948419	2.792418	2.808018
Hager	2.917219	2.932819	2.620817	3.042019	2.917219	2.886019
Ext.Tridiag.-1	2.886019	2.932819	2.761218	2.932819	2.730018	2.917219
Ext.ThreeExp.	2.979619	2.964019	2.605217	2.886019	3.042019	2.714417
Diag.4	2.901619	2.870418	2.574016	2.792418	2.948419	2.652017
Diag.5	2.792418	2.917219	2.574016	2.901619	3.026419	2.901619
Ext.Himm.	2.761218	2.714417	2.667617	2.964019	2.995219	2.854818
Ext.PSC1	2.932819	2.745618	2.714417	2.511616	3.026419	2.792418
FullHess.FH2	2.870418	2.948419	2.886019	2.839218	3.010819	2.948419
Ext.Bl.Diag.BD1	2.979619	2.886019	2.948419	2.886019	2.901619	2.542816
Quad.QF1	2.854818	2.870418	3.057620	2.964019	2.964019	2.886019
Ext.Quad.Pen.QP1	2.948419	2.808018	2.605217	2.964019	2.823618	2.542816
Quad.QF2	2.839218	2.620817	2.886019	2.979619	2.901619	2.683217
Ext.EP1	2.730018	2.402415	2.932819	2.698817	2.792418	2.652017
Ext.Tridiag.-2	2.683217	2.605217	2.839218	2.870418	2.886019	2.542816
Tridia	2.683217	2.511616	2.964019	2.823618	2.823618	2.511616
Arwhead	2.917219	2.995219	2.745618	2.823618	2.745618	2.012413
Dqdrtic	2.761218	2.995219	2.901619	2.823618	2.730018	2.589617
Quartc(Cute)	2.886019	2.776818	2.886019	2.776818	2.870418	2.839218
Dixon3dq(Cute)	2.808018	2.948419	2.948419	2.839218	2.917219	2.605217

Table 1.
$n = 1000$.

function	BZA	TTRMIL	MRMIL	MHS	MLS	hAO
Biggsb1(Cute)	2.792418	2.870418	2.870418	2.917219	2.979619	2.901619
Gen.quart.	2.917219	2.932819	2.464816	2.948419	2.808018	2.620817
Diag.7	2.574016	2.589617	2.870418	2.620817	3.026419	2.698817
Diag.8	2.730018	2.979619	2.839218	2.964019	2.792418	2.979619
Full Hess.FH3	2.948419	2.574016	2.698817	3.026419	2.636417	2.745618
Himmelbg	2.854818	3.010819	2.901619	2.854818	2.995219	2.730018
Ext.Pow.	2.901619	2.854818	2.761218	2.808018	2.870418	2.995219
Ext.Maratos	2.854818	2.948419	2.870418	2.995219	2.870418	2.917219
Ext.Cliff	2.964019	3.042019	2.854818	2.932819	2.886019	2.854818
Pert.quad.diag.	2.714417	3.104420	2.683217	2.964019	2.667617	2.901619
Ext.Wood	2.995219	2.932819	2.948419	2.948419	2.964019	2.948419

function	BZA	TTRMIL	MRMIL	MHS	MLS	hAO
Ext.Trigon.	2.792418	2.995219	2.839218	3.010819	2.995219	2.745618
Ext.Rosenbr.	2.964019	2.839218	2.948419	2.932819	2.995219	2.776818
Average	3.915625	3.928105	3.533423	3.868045	3.973345	3.722184

Table 2.
n = *1000*.

function	BZA	TTRMIL	MRMIL	MHS	MLS	hAO
Ext.Pen.	46.160696	46.831500	48.656712	66.284825	65.863622	63.695208
Pert.Quad.	48.375910	45.801894	52.307135	66.612427	66.113224	65.551620
Raydan1	12.994883	12.105678	13.759288	16.972909	16.598506	16.754507
Raydan2	1.170008	1.029607	1.076407	1.154407	1.092007	1.107607
Diag.1	8.845257	0.904806	1.076407	1.123207	1.170008	1.092007
Diag.2	8.658055	7.831250	7.924851	9.094858	10.358466	10.327266
Diag.3	8.361654	9.141659	8.673656	10.686068	10.358466	10.514467
Gen.Tridiag.-1	5.616036	5.382034	5.865638	6.021639	6.489642	6.364841
Hager	5.241634	4.851631	5.881238	6.286840	5.304034	6.021639
Ext.Tridiag.-1	5.007632	4.804831	5.740837	5.787637	6.224440	5.803237
Ext.ThreeExp.	4.882831	4.820431	5.522435	6.115239	6.333641	5.834437
Diag.4	4.929632	4.898431	5.179233	5.803237	6.177640	6.427241
Diag.5	5.694037	4.851631	5.538036	5.709637	5.896838	6.115239
Ext.Himm.	5.834437	5.116833	5.382034	6.099639	5.772037	6.411641
Ext.PSC1	5.023232	5.054432	5.163633	6.411641	6.115239	5.990438
FullHess.FH2	5.210433	4.929632	4.851631	6.068439	6.349241	6.349241
Ext.Bl.Diag.BD1	4.851631	5.007632	5.226033	6.364841	6.364841	5.569236
Quad.QF1	5.475635	5.662836	6.302440	6.177640	6.146439	6.286840
Ext.Quad.Pen.QP1	5.226033	5.163633	4.929632	6.130839	5.818837	5.943638
Quad.QF2	5.335234	4.836031	5.990438	6.084039	6.084039	6.084039
Ext.EP1	5.070032	5.038832	6.052839	6.115239	4.992032	6.177640
Ext.Tridiag.-2	4.851631	4.976432	4.851631	6.349241	5.990438	6.099639
Tridia	5.413235	4.820431	5.475635	5.569236	5.818837	6.021639
Arwhead	4.867231	4.882831	5.023232	6.099639	6.380441	6.177640
Dqdrtic	5.163633	4.945232	5.023232	5.428835	6.006038	5.850038
Quartc(Cute)	5.912438	5.350834	5.834437	5.787637	5.896838	6.193240
Dixon3dq(Cute)	5.428835	4.789231	5.163633	6.162039	5.616036	5.881238

Table 3.
n = *5000*.

where

$$\theta_k = 1 - \frac{\left(g_k^T d_{k-1}\right)^2}{\|g_k\|^2 \|d_{k-1}\|^2}.$$

But this new *CG* direction does not fulfill a descent condition, so further modification is made, namely, having in view [53], the authors [2] introduce

function	BZA	TTRMIL	MRMIL	MHS	MLS	hAO
Biggsb1(Cute)	5.148033	4.695630	5.413235	5.912438	6.052839	6.349241
Gen.quart.	5.288434	4.758031	5.023232	6.349241	6.052839	4.960832
Diag.7	5.163633	4.664430	5.054432	5.959238	6.193240	6.255640
Diag.8	5.787637	4.742430	4.898431	6.099639	5.600436	6.208840
Full Hess.FH3	5.444435	4.789231	5.569236	6.177640	6.162039	6.224440
Himmelbg	5.584836	6.130839	5.475635	5.475635	6.006038	5.912438
Ext.Pow.	5.569236	4.789231	4.773631	5.990438	5.772037	6.162039
Ext.Maratos	5.148033	5.740837	4.976432	6.021639	6.286840	6.130839
Ext.Cliff	5.943638	5.850038	4.976432	5.990438	5.304034	6.286840
Pert.quad.diag.	5.912438	6.427241	4.976432	6.318041	6.115239	6.068439
Ext.Wood	5.584836	5.647236	4.789231	6.255640	5.350834	6.021639
Ext.Trigon.	5.366434	5.709637	4.773631	6.115239	6.021639	5.787637
Ext.Rosenbr.	6.177640	5.319634	4.617630	6.333641	6.021639	6.021639
Average	7.79302995	7.327367	7.694749	9.287519525	9.206789	9.225899

Table 4.
$n = 5000$.

$$\overline{\beta}_k = \beta_k - \lambda\left(\frac{\|y_{k-1}\|\theta_k}{d_{k-1}^T y_{k-1}}\right)^2 g_k^T d_{k-1},$$

where $\lambda > \frac{1}{4}$ is a parameter. Also, the global convergence is proven under standard conditions.

It is worth to mention the next papers about this theme, which can be interesting [4, 14–17, 25–27].

3. Trust region methods

We remind that the basic idea of Newton method is to approximate the objective function $f(x)$ around x_k by using a quadratic model:

$$q^{(k)}(s) = f(x_k) + g_k^T s + \frac{1}{2} s_k^T G_k s,$$

where $g_k = \nabla f(x_k)$, $G_k = \nabla^2 f(x_k)$, and also use the minimizer s_k of $q^{(k)}(s)$ to set $x_{k+1} = x_k + s_k$.

Also, remind that Newton method can only guarantee the local convergence, i.e., when s is small enough and the method is convergent locally.

Further, Newton method cannot be used when Hessian is not positive definite.

There exists another class of methods, known as trust region methods. It does not use the line search to get the global convergence, as well as it avoids the difficulty which is the consequence of the nonpositive definite Hessian in the line search.

Furthermore, it produces greater reduction of the function f than line search approaches.

Here, we define the region around the current iterate:

$$\Omega_k = \{x : \|x - x_k\| \leq \Delta_k\},$$

where Δ_k is the radius of Ω_k, inside which the model is trusted to be adequate to the objective function.

Our further intention is to choose a step which should be the approximate minimizer of the quadratic model in the trust region. In fact, $x_k + s_k$ should be the approximately best point on the sphere:

$$\{x_k + s | \|s\| \leq \Delta_k\},$$

with the center x_k and the radius Δ_k.

In the case that this step is not acceptable, we reduce the size of the step, and then we find a new minimizer.

This method has the rapid local convergence rate, and that's the property of Newton method and quasi-Newton method, too, but the important characteristic of trust region method is also the global convergence.

Since the step is restricted by the trust region, this method is also called the restricted step method.

The model subproblem of the trust region method is

$$\min q^{(k)}(s) = f(x_k) + g_k^T s + \frac{1}{2} s^T B_k s, \tag{17}$$

$$\text{s.t.} \|s\| \leq \Delta_k, \tag{18}$$

where Δ_k is the trust region radius and B_k is a symmetric approximation of the Hessian G_k.

In the case that we use the standard l_2 norm $\|\cdot\|_2$, s_k is the minimizer of $q^{(k)}(s)$ in the ball of radius Δ_k. Generally, different norms define the different shapes of the trust region.

Setting $B_k = G_k$ in (17)–(18), the method becomes a Newton-type trust region method.

The problem by itself is the choice of Δ_k at each single iteration.

If the agreement between the model $q^{(k)}(s)$ and the objective function $f(x_k + s)$ is satisfactory enough, the value Δ_k should be chosen as large as it is possible. The expression $Ared_k = f(x_k) - f(x_k + s_k)$ is called the *actual reduction*, and the expression $Pred_k = q^{(k)}(0) - q^{(k)}(s_k)$ is called the *predicted reduction*; here, we emphasize that

$$r_k = \frac{Ared_k}{Pred_k}$$

measures the agreement between the model function $q^{(k)}(s)$ and the objective function $f(x_k + s)$.

If r_k is close to 0 or it is negative, the trust region is going to shrink; otherwise, we do not change the trust region.

The conclusion is that r_k is important in making the choice of new iterate x_{k+1} as well as in updating the trust region radius Δ_k. Now, we give the trust region algorithm.

Algorithm 1.3.1. *(Trust region method).*

Assumptions: x_0, $\overline{\Delta}$, $\Delta_0 \in (0, \overline{\Delta})$, $\varepsilon \geq 0$, $0 < \eta_1 \leq \eta_2 < 1$, and $0 < \gamma_1 < 1 < \gamma_2$.

Let $k = 0$.

Step 1. If $\|g_k\| \leq \varepsilon$, *then STOP.*

Step 2. Approximately solve the problem (17)–(18) for s_k.

Step 3. Compute $f(x_k + s_k)$ *and* r_k. *Set*

$$x_{k+1} = \begin{cases} x_k + s_k, & \text{if } r_k \geq \eta_1, \\ x_k, & \text{otherwise}. \end{cases}$$

Step 4. If $r_k < \eta_1$, *then* $\Delta_{k+1} \in (0, \gamma_1 \Delta_k)$.
If $r_k \in [\eta_1, \eta_2)$, *then* $\Delta_{k+1} \in (\gamma_1 \Delta_k, \Delta_k)$.
If $r_k \geq \eta_2$ *and* $\|s_k\| = \Delta_k$, *then* $\Delta_{k+1} \in \left[\Delta_k, \min\{\gamma_2 \Delta_k, \overline{\Delta}\}\right]$.
Step 5. Generate B_{k+1}, *update* $q^{(k)}$, *set* $k = k + 1$, *and go to Step 1.*

In Algorithm 1.3.1, $\overline{\Delta}$ is a bound for all Δ_k. Those iterations with the property $r_k \geq \eta_2$ (and so those for which $\Delta_{k+1} \geq \Delta_k$) are called *very successful iterations*; the iterations with the property $r_k \geq \eta_1$ (and so those for which $x_{k+1} = x_k + s_k$) are called *successful iterations*; and the iterations with the property $r_k < \eta_1$ (and so those for which $x_{k+1} = x_k$) are called *unsuccessful iterations*. Generally, the iterations from the two first cases are called *successful iterations*.

Some choices of parameters are $\eta_1 = 0, 01$, $\eta_2 = 0, 75$, $\gamma_1 = 0, 5$, $\gamma_2 = 2$, $\Delta_0 = 1$, and $\Delta_0 = \frac{1}{10} \|g_0\|$. The algorithm is insensitive to change of these parameters.

Next, if $r_k < 0, 01$, then Δ_{k+1} can be chosen from $(0.01, 0.5)\|s_k\|$ on the basis of a polynomial interpolation.

In the case of quadratic interpolation, we set

$$\Delta_{k+1} = \lambda \|s_k\|,$$

where

$$\lambda = \frac{-g_k^T s_k}{2\left(f(x_k + s_k) - f(x_k) - g_k^T s_k\right)}.$$

3.1 Convergence of trust region methods

Assumption 1.3.1 *(Assumption* A_0*).*

We assume that the approximations of Hessian $\{B_k\}$ *are uniformly bounded in norm and the level set* $L = \{x | f(x) \leq f(x_0)\}$ *is bounded, as well as* $f : \mathbb{R}^n \to \mathbb{R}$ *is continuously differentiable on L. We allow the length of the approximate solution* s_k *of the subproblem (17)–(18) to exceed the bound of the trust region, but we also assume that*

$$\|s_k\| \leq \tilde{\eta} \Delta_k,$$

where $\tilde{\eta}$ *is a positive constant.*

In this kind of trust region way of thinking, generally we do not seek an accurate solution of the subproblem (17)–(18); we are satisfied by finding a nearly optimal solution of the subproblem (17)–(18).

Strong theoretical as well as numerical results can be obtained if the step s_k, produced by Algorithm 1.3.1, satisfies

$$q_k(0) - q_k(s_k) \geq \beta_1 \|g_k\|_2 \min\left\{\Delta_k, \frac{\|g_k\|_2}{\|B_k\|_2}\right\}, \beta_1 \in (0, 1).$$

Theorem 1.3.1 *[47] Under Assumption* A_0, *if Algorithm 3.1 has finitely many successful iterations, then it converges to the first-order stationary point.*

Theorem 1.3.2 *[47] Under Assumption* A_0, *if Algorithm 3.1 has infinitely many successful iterations, then*

$$\liminf_{k\to\infty} \|g_k\| = 0.$$

In [44], it is emphasized that trust region methods are very effective for optimization problems and a new adaptive trust region method is presented. This method combines a modified secant equation with the *BFGS* update formula and an adaptive trust region radius, where the new trust region radius makes use of not only the function information but also the gradient information. Let $\hat{B}_k$ be a positively definite matrix based on modified Cholesky factorization [43]. Under suitable conditions, in [44] the global convergence is proven; also, the local superlinear convergence of the proposed method is demonstrated. Motivated by the adaptive technique, the proposed method possesses the following nice properties:

1. The trust region radius uses not only the gradient value but also the function value.

2. Computing the matrix $\hat{B}_k$ of the inverse and the value of $\|\hat{B}_k{}^{-1}\|$, at each iterative point x_k, is not required.

3. The computational time is reduced.

A modified secant equation is introduced:

$$B_{k+1}d_k = q_k, \tag{19}$$

where $q_k = y_k + h_k d_k, f_k = f(x_k)$, and $h_k = \frac{(g_{k+1}+g_k)^T d_k + 2(f_k - f_{k+1})}{\|d_k\|^2}$.

When f is twice continuously differentiable and B_{k+1} is generated by the *BFGS* formula, where $B_0 = I$, this modified secant Eq. (19) possesses the following nice property:

$$f_k = f_{k+1} - g_{k+1}^T d_k + \frac{1}{2} d_k^T B_{k+1} d_k,$$

and this property holds for all k.

Under classical assumptions, the global convergence of the method presented in [44] is also proven in this paper.

In [28], the hybridization of monotone and non-monotone approaches is made; a modified trust region ratio is used, in which more information is provided about the agreement between the exact and the approximate models. An adaptive trust region radius is used, as well as two accelerated Armijo-type line search strategies to avoid resolving the trust region subproblem whenever a trial step is rejected. It is shown that the proposed algorithm is globally and locally superlinearly convergent. In this paper trust region methods are denoted shortly by *TR*; it is emphasized that in *TR* method, having in view that the iterative scheme is

$$x_0 \in \mathbb{R}^n, x_{k+1} = x_k + s_k, k = 0, 1, \ldots,$$

and it often happens that s_k is an approximate solution of the following quadratic subproblem:

$$\min_{s\in\mathbb{R}^n,\ \|s_k\|\le\Delta_k} m_k(s) = g_k^T s + \frac{1}{2} s_k^T B_k s. \tag{20}$$

Performance of the *TR* methods is much influenced by the strategy of choosing the *TR* radius at each iteration. To determine the radius Δ_k, in the standard *TR* method, the agreement between $f(x_k + s)$ and $m_k(s)$ is evaluated by the so-called *TR* ratio ρ_k:

$$\rho_k = \frac{f(x_k) - f(x_k + s_k)}{m_k(0) - m_k(s_k)}.$$

When ρ_k is negative or a small positive number near to zero, the quadratic model is a poor approximation of the objective function. In such situation, Δ_k should be decreased and, consequently, the subproblem (20) should be solved again. However, when ρ_k is close to 1, it is reasonable to use the quadratic model as an approximation of the objective function. So, the step s_k should be accepted and Δ_k can be increased. Here, the authors use the modified version of ρ_k:

$$\overline{\rho}_k = \frac{R_k - f(x_k + s_k)}{P_k - m_k(s_k)},$$

where $R_k = \eta_k f_{l(k)} + (1 - \eta_k) f_k$, $\eta_k \in [\eta_{min}, \eta_{max}]$, $\eta_{min} \in [0, 1)$, and $\eta_{max} \in [\eta_{min}, 1]$. Also,

$$f_{l(k)} = \max_{0 \le j \le q(k)} \{f_{k-j}\}, f_i = f(x_i), q(0) = 0, 0 \le q(k) \le \min\{q(k-1) + 1, N\},$$

where $N \in \mathbb{N}$ which is originally used by Toint [48].

Something more about trust region methods can be found in [9, 18, 21, 22, 54].

4. Conclusion

The conjugate gradient methods and trust region methods are very popular now.

Many scientists consider these methods.

Namely, different modifications of these methods are made, with the aim to improve them.

Next, the scientists try to make not only new methods but also whole new classes of methods. For the specific values of the parameters, individual methods are distinguished from these classes. It is always more desirable to make a class of methods instead of individual methods.

Hybrid conjugate gradient methods are made in many different ways; this class of conjugate gradient methods is always actual.

Further, one of the contemporary trends is to use *BFGS* update in constructions of new conjugate gradient methods (e.g., see [46]).

Finally, let us emphasize that contemporary papers often use the Picard-Mann-Ishikawa iterative processes and they make the connection of these kinds of processes with the unconstrained optimization (see [29, 37, 38]).

Author details

Snezana S. Djordjevic
Faculty of Technology, University of Nis, Leskovac, Serbia

*Address all correspondence to: snezanadjordjevicle@gmail.com

References

[1] Adeleke OJ, Osinuga IA. A five-term hybrid conjugate gradient method with global convergence and descent properties for unconstrained optimization problems. Asian Journal of Scientific Research. 2018;**11**:185-194

[2] Amini K, Faramarzi P, Pirfalah N. A modified Hestenes-Stiefel conjugate gradient method with an optimal property. Optimization Methods and Software. In press. 2018. DOI: 10.1080/10556788.2018.1457150

[3] Andrei N. An unconstrained optimization test functions. Advanced Modeling and Optimization. An Electronic International Journal. 2008; **10**:147-161

[4] Andrei N. Acceleration of conjugate gradient algorithms for unconstrained optimization. Applied Mathematics and Computation. 2009;**213**:361-369

[5] Andrei N. 40 Conjugate Gradient Algorithms For Unconstrained Optimization. A Survey on Their Definition. ICI Technical Report No. 13/08, March 14, 2008

[6] Andrei N. A new three-term conjugate gradient algorithm for unconstrained optimization. Numerical Algorithms. 2015;**68**:305-321

[7] Baluch B, Salleh Z, Alhawarat A. A new modified three-term Hestenes-Stiefel conjugate gradient method with sufficient descent property and its global convergence. Hindawi Journal of Optimization. 2017;**2017**:1-13

[8] Beale EML. A derivative of conjugate gradients. In: Lootsma FA, editor. Numerical Methods for Nonlinear Optimization. London: Academic; 1972. pp. 39-43

[9] Curtis FE, Lubberts Z, Robinson DP. Concise complexity analyses for trust-region methods. Optimization Letters. 2018;**12**(8):1713-1724

[10] Dai YH, Kou CX. A nonlinear conjugate gradient algorithm with an optimal property and an improved Wolfe line search. SIAM Journal on Optimization. 2013;**23**(1):296-320

[11] Dai Y, Yuan Y. Alternate minimization gradient method. IMA Journal of Numerical Analysis. 2003;**23**: 377-393

[12] Dai YH, Yuan Y. A nonlinear conjugate gradient method with a strong global convergence property. SIAM Journal on Optimization. 1999;**10**: 177-182

[13] Dong X-L, Han D, Dai Z, Li L, Zhu J. An accelerated three-term conjugate gradient method with sufficient descent condition and conjugacy condition. Journal of Optimization Theory and Applications. 2018;**179**:944-961

[14] Du S, Chen M. A new smoothing modified three-term conjugate gradient method for l_1-norm minimization problem. Journal of Inequalities and Applications. 2018;**2018**(1):1-14. SpringerOpen

[15] Djordjević SS. New hybrid conjugate gradient method as a convex combination of FR and PRP methods. Univerzitet u Nišu. 2016;**30**(11): 3083-3100

[16] Djordjević SS. New hybrid conjugate gradient method as a convex combination of LS and CD methods. Univerzitet u Nišu. 2016;**31**(6): 1813-1825

[17] Djordjevic S. New hybrid conjugate gradient method as a convex combination of Ls and Fr methods. Acta Mathematica Scientia. 2019;**39**(1): 214-228

[18] El-Sobky B, Abo-Elnaga Y. A penalty method with trust-region mechanism for nonlinear bilevel optimization problem. Journal of Computational and Applied Mathematics. 2018;**340**:360-374

[19] Fletcher R, Reeves C. Function minimization by conjugate gradients. The Computer Journal. 1964;**7**:149-154

[20] Fletcher R. Practical methods of optimization. In: Unconstrained Optimization. Vol. 1. New York: John Wiley & Sons; 1987

[21] Gao J, Cao J. A class of derivative-free trust-region methods with interior backtracking technique for nonlinear optimization problems subject to linear inequality constraints. Journal of Inequalities and Applications. 2018; **2018**(1):108; https://www.ncbi.nlm.nih.gov/pmc/articles/PMC5942389/

[22] Gertz EM, Gill PE. A primal-dual trust region algorithm for nonlinear optimization. Mathematical Programming, Series B. 2004;**100**:49-94

[23] Hager WW, Zhang H. A new conjugate gradient method with guaranteed descent and an efficient line search. SIAM Journal on Optimization. 2003;**16**(1):170-192

[24] Hestenes MR, Stiefel EL. Methods of conjugate gradients for solving linear systems. Journal of Research of the National Bureau of Standards. 1952;**49**: 409-436

[25] Huang H, Lin S. A modified Wei-Yao-Liu conjugate gradient method for unconstrained optimization. Applied Mathematics and Computation. 2014; **231**:179-186

[26] Huang H et al. The proof of the sufficient descent condition of the Wei-Yao-Liu conjugate gradient method under the strong Wolfe-Powell line search. Applied Mathematics and Computation. 2007;**189**(2):1241-1245

[27] Jiang Z, Xu N. Hot spot thermal floor plan solver using conjugate gradient to speed up. Mobile Information Systems. 2018;**2018**:1-8

[28] Babaie-Kafaki S, Rezaee S. Two accelerated nonmonotone adaptive trust region line search methods. Numerical Algorithms. 2018;**78**:911-928

[29] Khan SH. A Picard-Mann hybrid iterative process. Fixed Point Theory and Applications. 2013;**2013**:69. DOI: 10.1186/1687-1812-2013-69

[30] Liu JK, Feng YM, Zou LM. Some three-term conjugate gradient methods with the inexact line search condition. Calcolo. 2018;**55**:16

[31] Liu Y, Storey C. Efficient generalized conjugate gradient algorithms, Part 1: Theory. Journal of Optimization Theory and Applications. 1991;**69**:129-137

[32] McGuire MF, Wolfe P. Evaluating a Restart Procedure for Conjugate Gradients, Report RC-4382. Yorktown Heights: IBM Research Center; 1973

[33] Narushima Y, Yabe H, Ford JA. A three-term conjugate gradient method with sufficient descent property for unconstrained optimization. SIAM Journal on Optimization. 2011;**21**: 212-230

[34] Nazareth L. A conjugate direction algorithm without line search. Journal of Optimization Theory and Applications. 1977;**23**:373-387

[35] Osadcha O, Marszaek Z. Comparison of steepest descent method and conjugate gradient method. In: CEUR Workshop Proceedings, SYSTEM 2017 - Proceedings of the Symposium for Young Scientists in Technology, Engineering and Mathematics. 2017; **1853**:1-4

[36] Perry A. Technical Note - A modified conjugate gradient algorithm.

Operations Research. 1978;**26**(6): 1073-1078

[37] Petrovic M, Rakocevic V, Kontrec N, et al. Hybridization of accelerated gradient descent method. Numerical Algorithms. 2018;**79**:769-786. DOI: 10.1007/s11075-017-0460-4

[38] Petrović MJ, Stanimirović PS, Kontrec N, Mladenov J. Hybrid modification of accelerated double direction method. Mathematical Problems in Engineering. 2018;**2018**:1-8

[39] Polak E, Ribiére G. Note sur la convergence de de directions conjugées. In: Rev. Francaise Informat Recherche Opertionelle, 3e Année. Vol. 16. 1969. pp. 35-43

[40] Polyak BT. The conjugate gradient method in extreme problems. USSR Computational Mathematics and Mathematical Physics. 1969;**9**:94-112

[41] Rivaie M, Mamat M, June LW, Mohd I. A new class of nonlinear conjugate gradient coefficient with global convergence properties. Applied Mathematics and Computation. 2012; **218**:11323-11332

[42] Rivaie M, Mamat M, Abashar A. A new class of nonlinear conjugate gradient coefficients with exact and inexact line searches. Applied Mathematics and Computation. 2015; **268**:1152-1163

[43] Schnabel RB, Eskow E. A new modified Cholesky factorization. SIAM Journal on Scientific Computing. 1990; **11**:1136-1158

[44] Sheng Z, Yuan G, CUI Z. A new adaptive trust region algorithm for optimization problems. Acta Mathematica Scientia. 2018;**38B**(2): 479-496

[45] Shengwei Y, Wei Z, Huang H. A note about Wyl's conjugate gradient method and its applications. Applied Mathematics and Computation. 2007; **191**:381-388

[46] Stanimirović PS, Ivanov B, Djordjević S, Brajević I. New hybrid conjugate gradient and Broyden–Fletcher–Goldfarb–Shanno Conjugate Gradient Methods. Journal of Optimization Theory and Applications. 2018;**178**:860-884

[47] Sun W, Yuan Y-X. Optimization theory and methods: Nonlinear programming. Springer Optimization and Its Applications. 2006;**1**

[48] Toint PhL. Nonmonotone trust region algorithms for nonlinear optimization subject to convex constraints. Mathematical Programming. 1997;**77**:69. DOI: 10.1007/BF02614518

[49] Wei Z, Yao S, Liu L. The convergence properties of some new conjugate gradient methods. Applied Mathematics and Computation. 2006; **183**(2):1341-1350

[50] Wolfe P. Convergence conditions for ascent methods. SIAM Review. 1969; **11**:226-235

[51] Wolfe P. Convergence conditions for ascent methods II: Some corrections. SIAM Review. 1969;**11**:226-235

[52] Wu G, Li Y, Yuan G. A three-term conjugate gradient algorithm with quadratic convergence for unconstrained optimization problems. Hindawi, Mathematical Problems in Engineering. 2018, Article ID: 4813030, 15 p. DOI: 10.1155/2018/4813030

[53] Yu GH, Guan LT, Chen WF. Spectral conjugate gradient methods with sufficient descent property for large scale unconstrained optimization. Optimization Methods and Software. 2008;**23**(2):275-293

[54] Zhang X, Zhang J, Liao L. An adaptive trust region method and its

convergence. Science in China, Series A: Mathematics. 2002;**45A**:620-631

[55] Zhang L, Zhou W, Li D. Some descent three-term conjugate gradient methods and their global convergence. Optimization Methods and Software. 2007;**22**(4):697-711

[56] Zhou W, Zhang L. A nonlinear conjugate gradient method based on the *MBFGS* secant condition. Optimization Methods and Software. 2006;**21**(5): 707-714

Chapter 3

What Determines EP Curve Shape?

Frank Xuyan Wang

Abstract

Propose use kurtosis divided by skewness squared as shape factor, and use the global or conditional minimum/maximum of this shape factor for selecting and differentiating distribution families. Semi-empirical formulas for that lower/upper bound are calculated for various distribution families, with the aid of Computer Algebra System, for fitting hard to match distributions. Previous studies show high CV distribution is hard to fit and simulate, this study extends that conclusion to cases with low CV but still hard to match EP curves, characterized by having shape factors close to 1. The maximal likelihood approach of distribution fit can tell us which distribution family is better suited for an empirical distribution, but the shape factor range information can tell us why a distribution cannot fit well, or is not suitable at all. So the shape factor, in a sense, determines the EP curve shape.

Keywords: Skewness, kurtosis, TVaR, shape factor, reinsurance, computer algebra system, Beta distribution, Kumaraswamy distribution, asymptotic expansion, GB2 distribution, numerical optimization, generalized hyperbolic distribution

1. Introduction

In reinsurance industry, losses for a contract are simulated and represented by the losses cumulative distribution function (CDF), survival, or quantile functions. The plots of these functions are called the EP curves with the following terminology [1]: for a given annual or aggregated loss, the probability of seeing annual loss exceeding that loss is the exceeding-probability (EP) or aggregate-exceeding-probability (AEP). The average of all annual losses exceeding that given loss is the AEP tail value at risk, called the AEP TVaR, or simply TVaR. The EP curve is represented by a table consisting of pairs of probability and loss. It is desirable to fit a parametric distribution to this table for a more succinct representation and more reasonable interpolations for values not in the table. Then which distribution family to use and what characteristics of the data are needed or determine the distribution are the questions to answer.

The (scaled) Beta distribution is widely used in reinsurance for fit loss or loss ratio, perhaps because the Beta distribution has only two parameters and very simple formulas for mean and standard deviation using these parameters, whose inverse function also has simple formulas, so that the two statistics of mean and standard deviation can be used to easily determine the parameters.

For about 85% of the perils, this approach works well, in the sense that the TVaR of the fitted distribution for quantile of interest, such as the 0.96, 0.99 or 0.996

quantile TVaR which is needed for pricing and risk monitoring, is close to a few percent of the original data TVaR. The remaining 15% perils, such as the North American Tornado Hail (NATH), Australia Wind Storm (AUWS), Hawaii Wind Storm (HIWS), and Mexico Earthquake (MXEQ), can have more than 10% deviations.

The maximum likelihood estimation method is a way to find alternative fitting distributions [2, 3]. Instead of finding approximations of the smoothed empirical distribution, we optimize an objective function whose optimum solution gives us the candidate distribution form. Suppose the annual losses x_i occurred n_i times in our observation; to find a probability function that gives probabilities p_i for these losses, we just maximize the objective function $\prod_i {p_i}^{n_i}$. It is easily seen that for the optimum solution we have $\frac{p_i}{p_j} = \frac{n_i}{n_j}$: the relative occurring frequency is maintained in the probability function. In the objective function, if we replace the p_i by a power function of p_i, the conclusion still holds, but not if we use a logarithm or exponential function.

While the maximum likelihood approach works well for many perils and identifies a few best fitted distribution families (Mathematica has more than 200 distribution families that can be used for extensive searches), it did not work for the NATH peril. The NATH has {Mean, StandardDeviation, Skewness, Kurtosis, 0.99TVaR} = {7418611.10904006, 9517336.93024634, 5.99378199789956, 65.8901734355745, 68867612.8345741}.

This is not contradictory to the maximum likelihood principal, since in any implementation, only known forms of the probability density function (PDF) and as-small-as-possible numbers of parameters can be used. To overcome this limitation, we need to look into the particularity of those distributions and come up with or select more suitable function forms for the PDF or CDF. In [4] it is found that a high coefficient of variation (CV) distribution is hard to fit or simulate. But the NATH has a small CV of 1.28. The skewness and kurtosis alone also not differentiate them from other distributions.

Trial and error found the empirical rule that these hard distributions have small values of kurtosis divided by skewness squared, **Table 1**. This finding prompted us for the study of the property of kurtosis/skewness^2 (K/S^2), henceforth will be called the shape factor (SF).

Numerical optimization or solution will be our primary tool for this SF study. Analytical deduction, symbolic algebra, and symbolic limit from computer algebra system (CAS) Mathematica will be another major tool, as well as Mathematica's plot functions. Those plots can help reveal the patterns or tendencies of functions. The found pattern can in turn aid in taking special directional/constraint limit or substitutions in CAS to get the analytical formula for SF bound when it is possible.

The overall lower bound we find of SF is presented in Section 2, through the triple analytical, graphical, and numerical methods. Followed by in-detail studies of SF of various selected distribution families, which are either widely used in practice,

Peril	CV	Shape factor
NATH	1.283	1.834
AUWS	5.711	1.260
HIWS	4.678	1.238
MXEQ	3.930	1.878

Table 1.
Numerical characteristics of a few hard to fit and simulate perils.

such as the Beta distribution in Section 3 and the generalized Gamma distribution in Section 6, or is most simple to simulate, such as the Kumaraswamy distribution in Section 4. The most inclusive distribution, BetaPrime distribution, is in Section 5, for which we do not get an analytical formula, so the empirical formula for SF lower bound is provided. Some distributions that have wide matching capabilities, but for the NATH may have fitted distribution facing numerical difficulties, such as the Fleishman distribution, whose fit has non-monotonically increasing polynomial form and hence is hard to solve for inverse CDF, are only briefly mentioned in Section 7. The top distribution found through maximum likelihood fit, the generalized hyperbolic distribution (GH), even with the most complex PDF, has unexpectedly simple and beautiful analytical formulas for SF lower bound; the results are in the final Section 8. All our studies will focus on SF bound deductions and applications.

2. Lower bound of the shape factor

For a random variable f with mean m_f, the following characteristics are defined:

- Moment (M), $M[r] \equiv \int f^r d\mu, r > 0,$
- Absolute Moment (AM), $AM[r] \equiv \int |f|^r d\mu, r > 0,$
- Central Moment (CM), $CM[r] \equiv \int (f - m_f)^r d\mu, r > 0,$
- Absolute Central Moment (ACM), $ACM[r] \equiv \int |f - m_f|^r d\mu, r > 0,$
- Skewness (S), $S \equiv \frac{CM[3]}{CM[2]^{\frac{3}{2}}},$
- Kurtosis (K), $K \equiv \frac{CM[4]}{CM[2]^2},$
- Shape Factor (SF), $SF \equiv \frac{K}{S^2} = \frac{CM[4]*CM[2]}{CM[3]^2}.$

We can prove by Hölder inequality (https://en.wikipedia.org/wiki/Hölder's_inequality) that.

$SF \geq 1:$

$$|\int (f - m_f)^3 d\mu| \leq \int |f - m_f|^3 d\mu = \int |f - m_f|^2 |f - m_f|^1 d\mu \tag{1}$$

$$\leq \left(\int |f - m_f|^4 d\mu\right)^{\frac{1}{2}} \left(\int |f - m_f|^2 d\mu\right)^{\frac{1}{2}}. \tag{2}$$

A better inequality $K \geq S^2 + 1$ is proved in [5–7]. But by Hölder inequality we can also know that $\frac{ACM[4]*ACM[2]}{ACM[3]^2} = 1$ iff f is constant: if f is not constant, the shape factor must be larger than the lower bound 1.

The contribution to $SF > 1$ plausibly comes from two parts: Eq. (1) due to symmetry, the more symmetric the distribution, the larger the contribution to SF, or conversely, the smaller the SF, the more asymmetric the distribution; and Eq. (2) due to ACM convexity or steepness, the steeper the PDF, the smaller the SF.

This property of the shape factor identified our exceptional perils as possessing very steep and asymmetric PDF whose SF are small.

2.1 Are there better definitions of shape factor?

To measure the steepness or the convexity, we can get similar inequality to Eq. (2) by Hölder inequality for absolute moment:

$$\frac{AM[4]*AM[2]}{AM[3]^2} \geq 1 \text{ and } \frac{AM[3]*AM[1]}{AM[2]^2} \geq 1.$$

From absolute central moment define:

$$SF1 \equiv \frac{ACM[4]*ACM[2]}{ACM[3]^2} \geq 1 \text{ and } SF2 \equiv \frac{ACM[3]*ACM[1]}{ACM[2]^2} \geq 1.$$

For nonnegative random variables such as the reinsurance contract loss distribution, use the following inequality for moment:

$$\frac{M[4]*M[2]}{M[3]^2} \geq 1 \text{ and } \frac{M[3]*M[1]}{M[2]^2} \geq 1.$$

From another application of Hölder inequality, we get yet other measures of convexity from absolute central moment:

$$ACM[r] \leq ACM[s]^{\frac{r}{s}}, \textit{ where } 0 < r < s,$$

$SF3[r] \equiv \frac{ACM[r]}{ACM[1]^r} \leq 1$, *where* $0 < r < 1$ and $SF3[s] \equiv \frac{ACM[s]}{ACM[1]^s} \geq 1$, *where* $s > 1$.
Similar definition from absolute moment:

$$AM[r] \leq AM[s]^{\frac{r}{s}}, \textit{ where } 0 < r < s,$$

$SF4[r] \equiv \frac{AM[r]}{AM[1]^r} \leq 1$, *where* $0 < r < 1$ and $SF4[s] \equiv \frac{AM[s]}{AM[1]^s} \geq 1$, *where* $s > 1$.

Checking against $NormalDistribution[\mu, \sigma]$, we see their minimum based on absolute moment: $\frac{AM[4]*AM[2]}{AM[3]^2}$, $\frac{AM[3]*AM[1]}{AM[2]^2}$, and $SF4[2] = \frac{AM[2]}{AM[1]^2}$, are all 1, but that by absolute central moment are not: min SF1 = 1.1781, min SF2 = 1.27324, min SF3 [2] =1.5708. Moreover, the convex index SF1, SF2, and SF3 out of absolute central moment are shift invariant besides the scale transformation invariant of the random variable, so they are preferred over the ones based on absolute moment.

The only case in favor of $\frac{M[4]*M[2]}{M[3]^2}$ and $\frac{M[3]*M[1]}{M[2]^2}$ is when the numerical calculation error with extreme parameters arrive at negative kurtosis, then the calculated *SF* are meaningless (An example of *BesselK* function inaccuracy brings about negative kurtosis for generalized hyperbolic distribution can be found in [8]).

Even though both *SF* and *SF*1 are invariant under linear transformation of the distribution, and both measure the convexity, $SF \geq SF1$ can additionally measure the asymmetry, combining these two into one quantity. Since most distributions in reinsurance are not symmetric, *SF* is preferred over *SF*1. That only *SF* measured both asymmetry and convexity, while the others cannot, can be seen from **Figure 1**, for the case of exponential distribution family with PDF $e^{-x^n} n x^{-1+n}, x \in (0, \infty), n > 0$, which is $WeibullDistribution[n, 1]$ or $GammaDistribution[1, 1, n, 0]$, where only *SF* has a nontrivial interior global minimum.

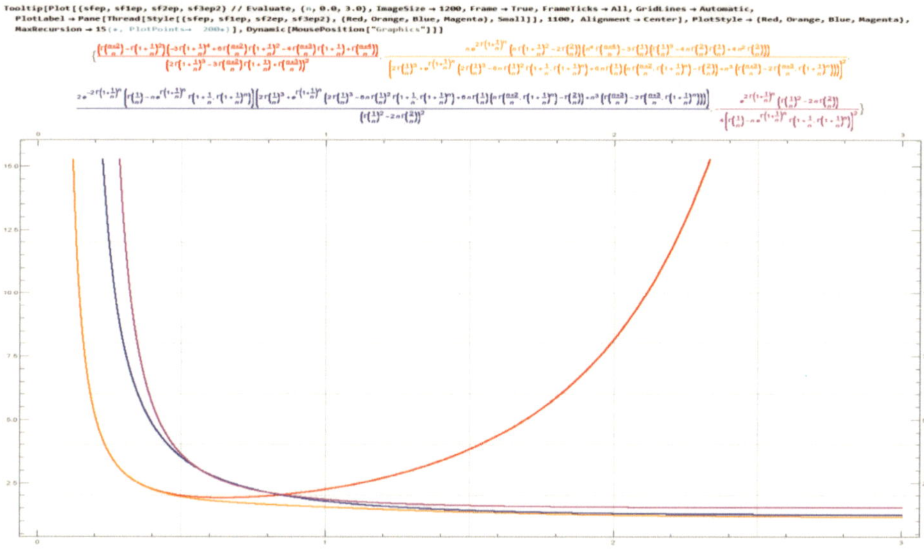

Figure 1.
SF SF1 SF2 SF3[2] plot of exponential distribution $e^{-x^n}nx^{-1+n}$. *The horizontal axis is the order of the exponential and the vertical axis is shape factors values.*

An intuitive reason for why using shape factor *SF* in favor of skewness and kurtosis alone is provided by studying the simple example power distribution family with PDF $\frac{n+1}{n}x^{\frac{1}{n}}, x \in [0,1], n < -1 || n > 0$ (or *BetaDistribution*$[1/n+1, 1]$). This distribution family has the largest value of skewness and kurtosis, and at the same time the smallest shape factor *SF* when n turns to -1, where the PDF is the steepest, but the skewness and kurtosis take the indistinguishable value of infinity. In comparison, the shape factor *SF* takes the finite and distribution family specific value of 1.125. The shape factor *SF* thus makes meaning out of the meaningless infinities.

2.2 Alternative way of defining shape factor for symmetric distribution

For symmetric distribution, $CM[3] = 0$, our *SF* will be indiscriminately infinity. We can now employ *SF*1 in place of *SF*. Other measures from *ACM* such as *SF*2 and *SF*3 may also be candidates. From the *SF*3 plot **Figure 2** of *NormalDistribution*$[\mu, \sigma]$ we see that $\min_{0<r<1} \mathrm{SF3}[r] = 0.919824$. The lower the value of *SF*3[2], the higher the $\min_{0<r<1} \mathrm{SF3}[r]$. We can use either *SF*3[2] or $\min_{0<r<1} \mathrm{SF3}[r]$ as a shape factor for symmetric distribution to describe the convexity of the *ACM* curve. The second measure

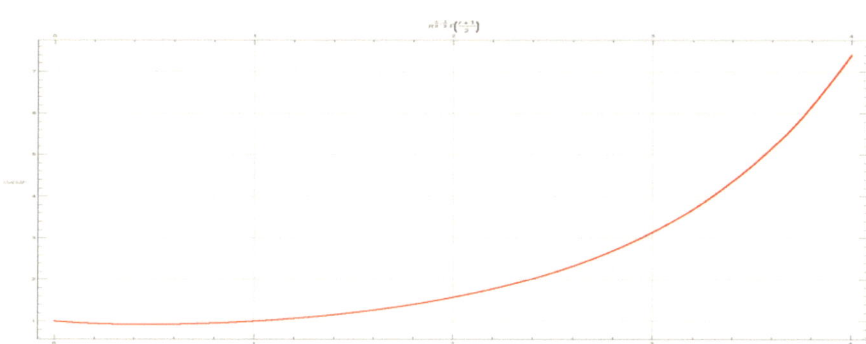

Figure 2.
*SF*3 *plot of Normal distribution. The horizontal axis is the order r of the power and the vertical axis is SF*3$[r]$.

has the merit of independence to the power order *r*, by additional efforts of numerical minimization. For our power distribution family, the maximum of the minimum is: $\max_{n>0} \min_{0<r<1} SF3[r] = 0.942085$, higher than the Normal distribution family.

When all *SF*, *SF*1, and *SF*2 are available, however, we will prefer *SF* to *SF*1 and *SF*2 since its dependency on parameters show simpler patterns than the other two; this can be shown from their contour plots for Beta distribution **Figures 3–5**, where *SF* contours are almost lines.

2.3 Lower bound of *SF* for well-known distributions

Using numerical optimization [9, 10], for most of the top-fitted distributions from the maximum log likelihood approach, we get the minimum *SF* values, with distribution definition in [11] whose naming and parameterization for probability distributions will be used throughout this chapter, in **Table 2**.

From this table, we know that most of the distributions are not able to describe NATH since NATH has *SF* 1.834. More involved numerical integration and optimization also eliminated the Beckmann Distribution [12], with admissible SF range

Figure 3.
*SF*1 *contour plot of Beta distribution. The horizontal axis is α and the vertical axis is β.*

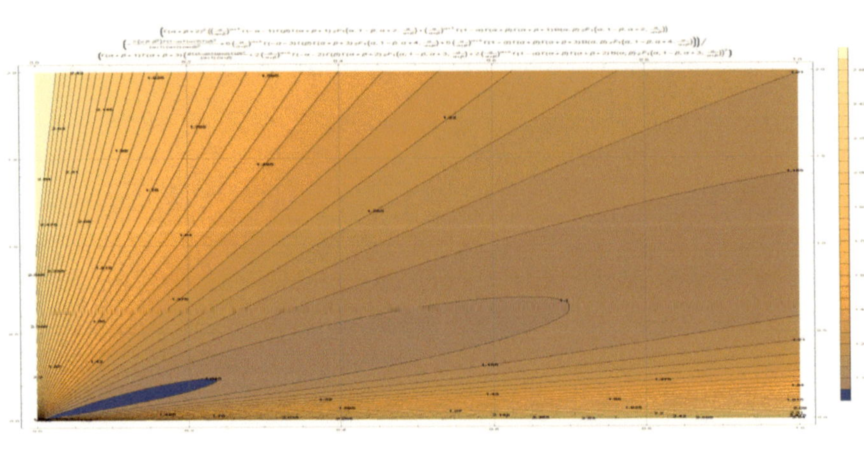

Figure 4.
*SF*2 *contour plot of Beta distribution. The horizontal axis is α and the vertical axis is β.*

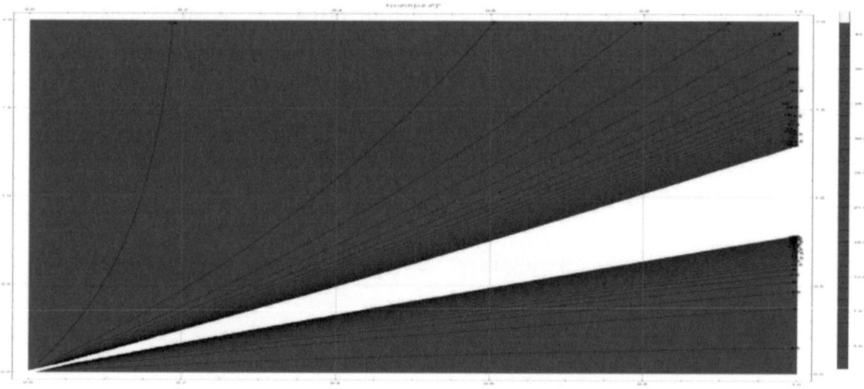

Figure 5.
SF contour plot of Beta distribution. The horizontal axis is α and the vertical axis is β.

Distribution	Min *SF*	Location of the Min
FrechetDistribution[α, β, μ]	2.9555	$\alpha \to 7.9305$
ExtremeValueDistribution[α, β]	4.15843	any α, β
MaxStableDistribution[μ, σ, ξ]	1.91227	$\xi \to -1.55970090120176$
InverseGaussianDistribution[μ, λ, θ]	1.5	$\lambda/\mu \to 0$
SkewNormalDistribution[μ, σ, α]	3.90603	$\alpha \to \infty$
ExpGammaDistribution[κ, θ, μ]	2.25	$\kappa \to 0$
BirnbaumSaundersDistribution[α, λ]	1.63481	$\alpha \to \infty$
MeixnerDistribution[a, b, m, d]	1.5	$d \to 0, b \to \pm\pi$

Table 2.
Lower bound of SF for some well-known distributions.

3.63–8.16, being the top four-parameter-distribution in another distribution fit case study that has SF 4.58.

The Alpha-Skew-Normal Distribution from [13] has minimum *SF*4.95061 when α is 2.07764, from its proposition 2.3, is thus also not eligible for NATH.

The global lower bound of *SF* for parametric distribution can be used to filter out those distributions whose values are larger than the losses data *SF*, so that we can focus on distributions that do not violate the bound. In the following sections we will study typical distribution SF bound, beginning with the Beta distribution.

3. Beta distribution

Regardless of the fact that multitude distribution types have been used for the frequency and severity distribution of individual contract losses, the aggregated portfolio losses for the majority of perils can be fitted by a compound Poisson distribution with Beta distribution as the severity, somehow an attest of its prevalence. Beta distribution has min $SF = 1.0$, so we need an in-detail study of why it cannot fit NATH.

When matching a *BetaDistribution*[α, β] for skewness 5.99378 and kurtosis 65.8902, we must have $\beta < 0$. When matching a Beta distribution for CV(=std/mean, the standard deviation divided by the mean) 1.2829 and either skewness

5.99378 or kurtosis 65.8902, we must have either both $\alpha < 0$ and $\beta < 0$ or at least one of α or β less than 0. Since CV, skewness, and kurtosis are scale invariant, so no scaled Beta distribution can at the same time match any two of the three statistics CV, skewness, and kurtosis.

3.1 Minimum shape factor for given CV

Using Mathematica, we can solve the parameter α and β by cv and std for $BetaDistribution[\alpha, \beta]$:

$$\alpha \to \frac{\mathrm{cv} - \mathrm{std} - \mathrm{cv}^2\mathrm{std}}{\mathrm{cv}^3}, \beta \to \frac{\mathrm{cv}^2 - 2\mathrm{cvstd} - \mathrm{cv}^3\mathrm{std} + \mathrm{std}^2 + \mathrm{cv}^2\mathrm{std}^2}{\mathrm{cv}^3\mathrm{std}}.$$

Since $\alpha > 0$, we must have:

$$\mathrm{std} < \frac{\mathrm{cv}}{1 + \mathrm{cv}^2},$$

or

$$\frac{1 - \sqrt{1 - 4\mathrm{std}^2}}{2\mathrm{std}} < \mathrm{cv} < \frac{1 + \sqrt{1 - 4\mathrm{std}^2}}{2\mathrm{std}}.$$

We also know std must be between 0 and 0.5 for these solutions to exist. By computer-aided exploration through contour plot, we can find the location of the std where *SF* takes minimum for a given cv.

The overall observation is that when cv < 1, *SF* approaches infinity in the middle value of std, and decreases when deviating from it. When cv > 1, *SF* approaches its minimum in the middle value of std and increases when departing. Together with the fact that std has an allowable upper bound of cv/(1 + cv^2) and lower bound of 0, the minimum of *SF* must be attained either at the global extreme where the derivative of *SF* with respect to std is zero or at the two boundaries when cv > 1, and attained at the two boundaries when cv < 1.

Using Mathematica to take the derivative of the shape factor with respect to std to find the std where shape factor attained extreme values, and solving it for the intersection with std upper and lower bound, we know the minimal shape factor for Beta distribution for a given CV when CV is below 0.707107 or above 2.48239 (intersecting std upper bound) is attained at std upper bound $\frac{\mathrm{cv}}{1+\mathrm{cv}^2}$ with value:

$$\min_{0 < std < \frac{cv}{1+cv^2}} SF = \frac{1 - cv^2 + cv^4}{(-1 + cv^2)^2}, when\ cv < 0.707107 \| cv > 2.48239. \tag{3}$$

When CV is between 0.707107 and 1.024766 (intersecting std lower bound) the minimal shape factor is attained at std lower bound 0 with value:

$$\min_{0 < std < \frac{cv}{1+cv^2}} SF = 1.5 + \frac{0.75}{cv^2}, when\ 0.707107 \leq cv \leq 1.024766. \tag{4}$$

When CV is between 1.024766 and 2.48239, the minimum *SF* is attained at std that is the zero derivative points of the shape factor. The piecewise curve plot of the minimum SF for given CV is in **Figure 6**. The formula for the central piece, minshape, is given in **Figure 7** which is too complex for manual derivation without the aid of computer algebra system.

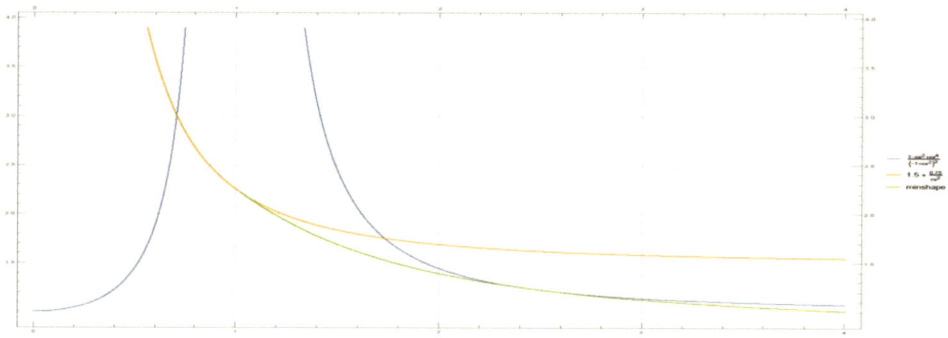

Figure 6.
Plot of Beta distribution min shape factor for given cv.

Figure 7.
Formula for minshape obtained using Mathematica.

From the curve we know when CV = 1.28, the minimal shape factor is 1.88, larger than 1.83 of NATH. In the best effort to match the input, we may elect to relax CV, for example, to 1.3, then the minimum shape factor is 1.85. With the constraint of a given CV, the minimum shape factor of the Beta Distribution may be significantly larger than its global minimum 1, so that it cannot attain to the wanted *SF* value.

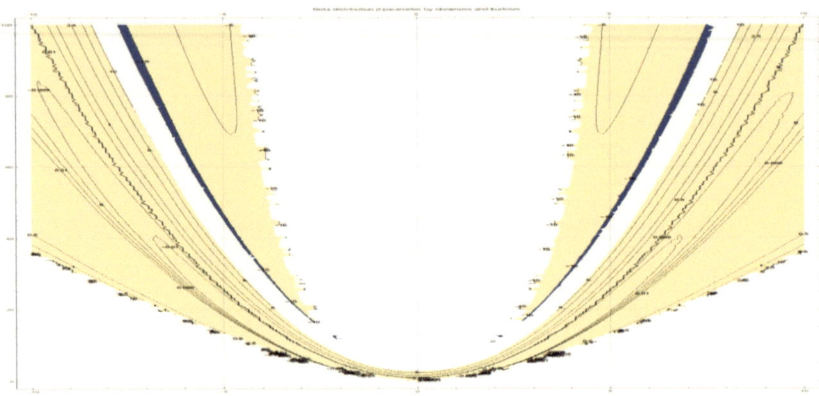

Figure 8.
Contour plot of Beta distribution β parameter. The horizontal axis is the skewness and the vertical axis is the kurtosis.

3.2 Shape factor range for given skewness

By solving Beta distribution parameters α and β through skewness *sk* and kurtosis *kt*, and examining the contour plot of β, we can see the allowable region is bound by two parabolas, **Figure 8**.

For a fixed skewness, α is monotonic increasing with respect to kurtosis; on the other hand, β has a singular point in some kurtosis, below that kurtosis is positive and monotonic increasing(in the region where α is positive), **Figure 9**.

Solving for that singular point we get the permissible kurtosis upper bound $3 + \frac{3}{2}\, sk^2$, and solve for $\beta = 0$ get the permissible kurtosis lower bound $1 + sk^2$.

Observe that the upper bound is when β turns to infinity, we can also get a simpler derivation of the upper bound by representing skewness and shape factor in α and β, letting $\beta \to \infty$, and then eliminating α to get shape factor as a function of skewness (Mathematica cannot solve equation for skewness which includes square root expression, we get around that by solving equation for the square of skewness, and then abandoning the negative solution introduced by this square).

A third way of more tedious calculation is through solving α by skewness and β, substituting the real solution into shape factor, and then take the limit for $\beta \to \infty$.

All three methods get the same upper bound of $SF = \frac{3}{2} + \frac{3}{sk^2}$.

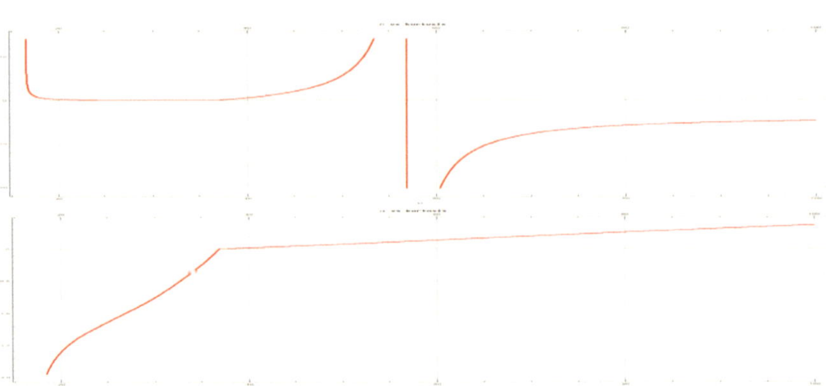

Figure 9.
Plots of Beta distribution β parameter and α parameter vs. kurtosis for a given skewness 5.99378.

So for Beta distribution, the allowable region of skewness and kurtosis is bound below by kurtosis = skewness^2 + 1 where $\beta \to 0$, and above by kurtosis = 3 + 1.5*skewness^2 where $\beta \to \infty$:

$$1+\frac{1}{S^2}\leq SF\leq 1.5+\frac{3}{S^2}. \tag{5}$$

For the given skewness of 5.99378 of NATH, the maximum allowable kurtosis is 56.88813, less than the wanted 65.8902. So NATH cannot be fitted by any affine transformation of Beta distribution, certifying NATH as a trying case for distribution fitting. We will use it to test many of the well-known distributions in later sections. We also see surprisingly that unlike many of the other distribution families whose shape factors are too high, the Beta distributions have the shape factor range too low, or too close to 1. This suggests us to search for distributions with shape factors ranges in between.

4. Kumaraswamy distribution

Using the same approach as in the Beta distribution, we first study the skewness and kurtosis tendency of *KumaraswamyDistribution*$[\alpha,\beta]$ [14], since the latter tested

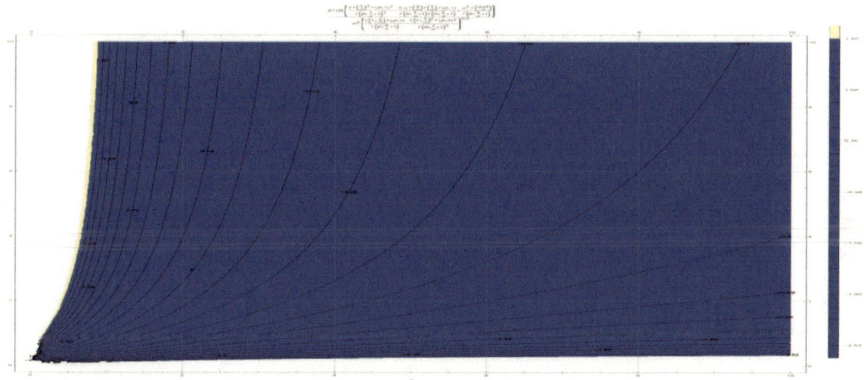

Figure 10.
Contour plot of Kumaraswamy distribution skewness.

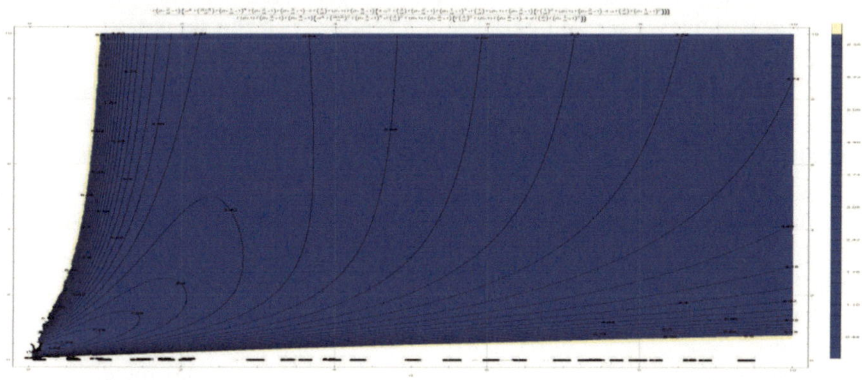

Figure 11.
Contour plot of Kumaraswamy distribution kurtosis.

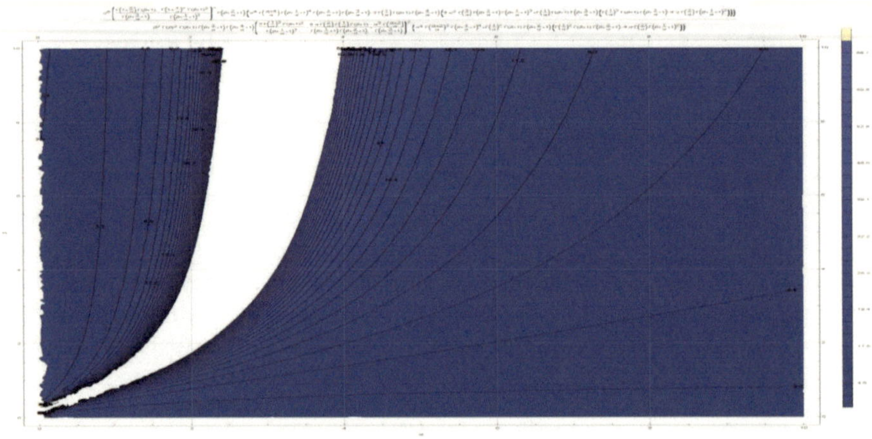

Figure 12.
Contour plot of Kumaraswamy distribution shape factor.

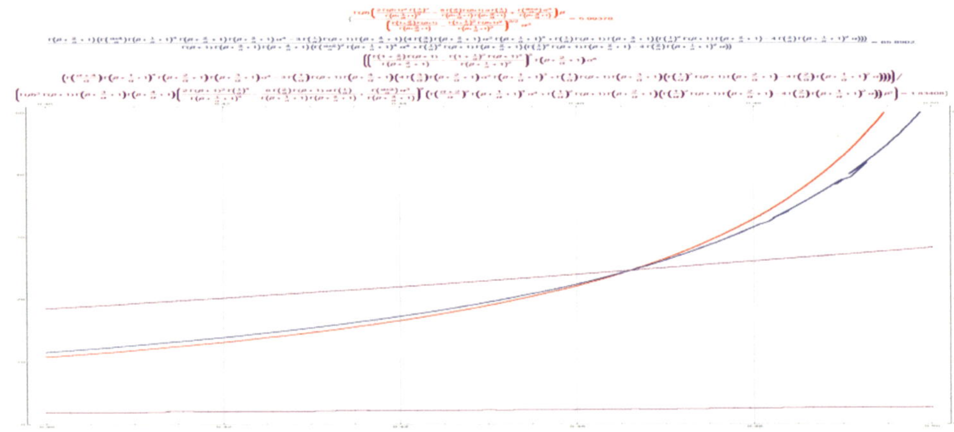

Figure 13.
Contour plot of Kumaraswamy distribution skewness, kurtosis, and shape factor for given values 5.99, 65.89, and 1.83. The horizontal axis is the α *parameter and the vertical axis is the* β *parameter.*

to be a better choice in our experiment and is also the easiest for simulation, **Figures 10–12**; and then study the SF bound for given skewness.

From these plots, we see an overall rough tendency of the skewness, kurtosis and shape factor. For a given α, the shape factor converges to a finite limit when $\beta \to \infty$. For a given skewness or a given kurtosis, there exists a maximum allowable α that is arrived when $\beta \to \infty$. In the parameters space of (α,β), for a given α, the kurtosis is increasing with respect to β in the top left portion where the skewness is positive, and decreasing in the right bottom portion where the skewness is negative. And in the parameters space of (α,β), for a given α, the shape factor is decreasing with respect to β in the top left portion where the skewness is positive, and increasing in the right bottom portion where the skewness is negative. But we will see later that the tendencies are more delicate than the monotonicity shown through visual observation.

Combining the tendency of shape factor and the contour plot for given skewness, kurtosis, and shape factor as in **Figure 13**, we may guess that for a given positive skewness, when α turn to its upper limit and β turn to infinity, the shape

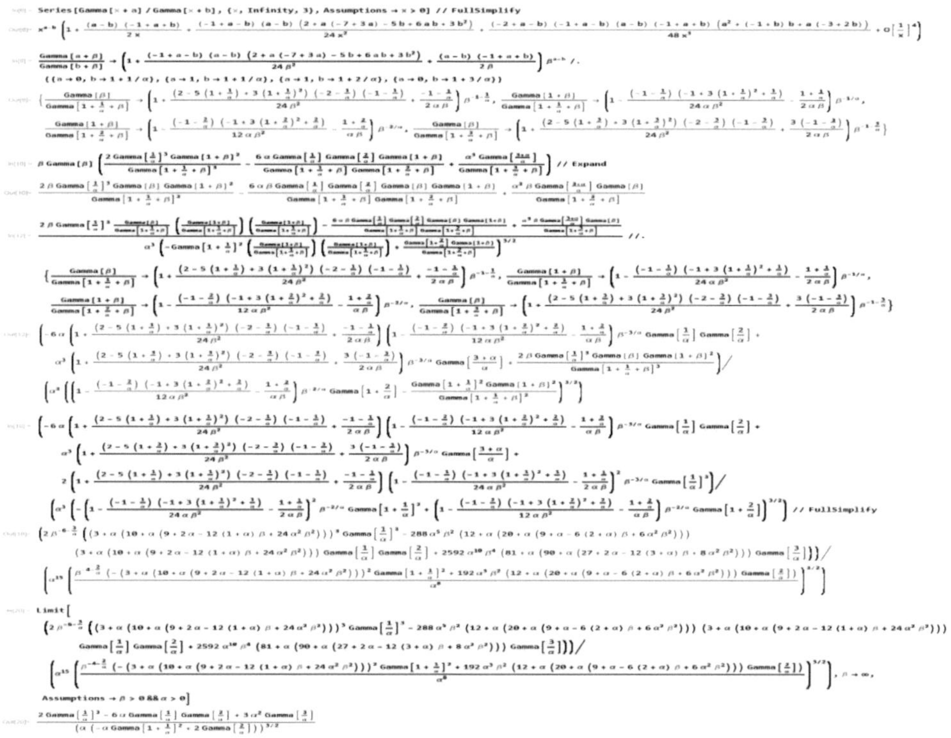

Figure 14.
Derivation of Kumaraswamy distribution skewness upper bound for given α.

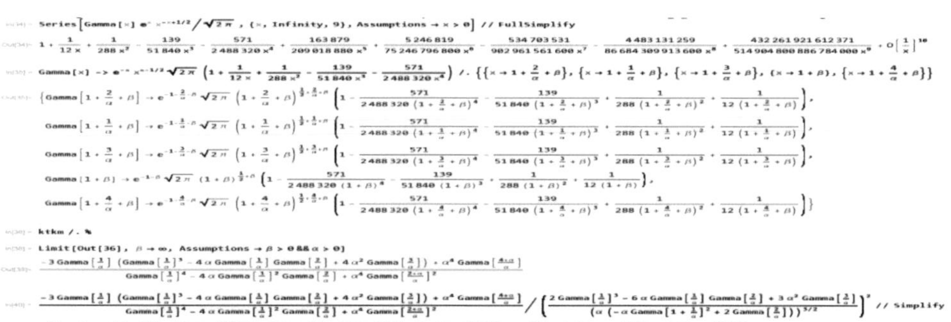

Figure 15.
Derivation of Kumaraswamy distribution kurtosis upper bound for given α and shape factor boundary value for given α when β → ∞.

factor will converge to its minimum. We use Mathematica to calculate the asymptotic expansion of the Gamma function and the quotient of Gamma function at infinity for orders up to 4 or 2, take the symbolic limit for $\beta \to \infty$, to get these boundary values, **Figures 14** and **15**.

We thus have a simple formula for boundary value of Kumaraswamy distribution shape factor:

$$\lim_{\beta\to\infty} S = \frac{2\mathrm{Gamma}\left[\frac{1}{\alpha}\right]^3 - 6\alpha\mathrm{Gamma}\left[\frac{1}{\alpha}\right]\mathrm{Gamma}\left[\frac{2}{\alpha}\right] + 3\alpha^2\mathrm{Gamma}\left[\frac{3}{\alpha}\right]}{\left(\alpha\left(-\alpha\mathrm{Gamma}\left[1+\frac{1}{\alpha}\right]^2 + 2\mathrm{Gamma}\left[\frac{2}{\alpha}\right]\right)\right)^{3/2}}, \tag{6}$$

$$\lim_{\beta\to\infty} K = \frac{-3\text{Gamma}\left[\frac{1}{\alpha}\right]\left(\text{Gamma}\left[\frac{1}{\alpha}\right]^3 - 4\alpha\text{Gamma}\left[\frac{1}{\alpha}\right]\text{Gamma}\left[\frac{2}{\alpha}\right] + 4\alpha^2\text{Gamma}\left[\frac{3}{\alpha}\right]\right) + \alpha^4\text{Gamma}\left[\frac{4+\alpha}{\alpha}\right]}{\text{Gamma}\left[\frac{1}{\alpha}\right]^4 - 4\alpha\text{Gamma}\left[\frac{1}{\alpha}\right]^2\text{Gamma}\left[\frac{2}{\alpha}\right] + \alpha^4\text{Gamma}\left[\frac{2+\alpha}{\alpha}\right]^2}, \tag{7}$$

$$\lim_{\beta\to\infty}\frac{K}{S^2} = \frac{\alpha^3\left(-\alpha\text{Gamma}\left[1+\frac{1}{\alpha}\right]^2 + 2\text{Gamma}\left[\frac{2}{\alpha}\right]\right)^3\left(-3\text{Gamma}\left[\frac{1}{\alpha}\right]\left(\text{Gamma}\left[\frac{1}{\alpha}\right]^3 - 4\alpha\text{Gamma}\left[\frac{1}{\alpha}\right]\text{Gamma}\left[\frac{2}{\alpha}\right] + 4\alpha^2\text{Gamma}\left[\frac{3}{\alpha}\right]\right) + \alpha^4\text{Gamma}\left[\frac{4+\alpha}{\alpha}\right]\right)}{\left(2\text{Gamma}\left[\frac{1}{\alpha}\right]^3 - 6\alpha\text{Gamma}\left[\frac{1}{\alpha}\right]\text{Gamma}\left[\frac{2}{\alpha}\right] + 3\alpha^2\text{Gamma}\left[\frac{3}{\alpha}\right]\right)^2\left(\text{Gamma}\left[\frac{1}{\alpha}\right]^4 - 4\alpha\text{Gamma}\left[\frac{1}{\alpha}\right]^2\text{Gamma}\left[\frac{2}{\alpha}\right] + \alpha^4\text{Gamma}\left[\frac{2+\alpha}{\alpha}\right]^2\right)}. \tag{8}$$

Its plot **Figure 16** has two branches, the dividing point is $\alpha \to 3.602349425719043$ where the skewness is zero, and below it is mainly the positive skewness region while above it is the negative skewness region.

The minimum value at the left branch of **Figure 16** is 1.91227 and arrived at $\alpha = 0.641149$. When $\alpha > 1000$ the numerical value for that boundary can be negative and is thus unreliable. The value 1.91227 is not the global minimum of the shape factor: for $\alpha = 0.641149$ the shape factor plot **Figure 17** with respect to β decreases first, at the point 10.6095 arriving at the minimum value of 1.80935, and increasing after the point 10.6095.

In principle, the extreme value of the shape factor for a given skewness will arrive either at the upper boundary where $\beta \to \infty$ or at the lower boundary where

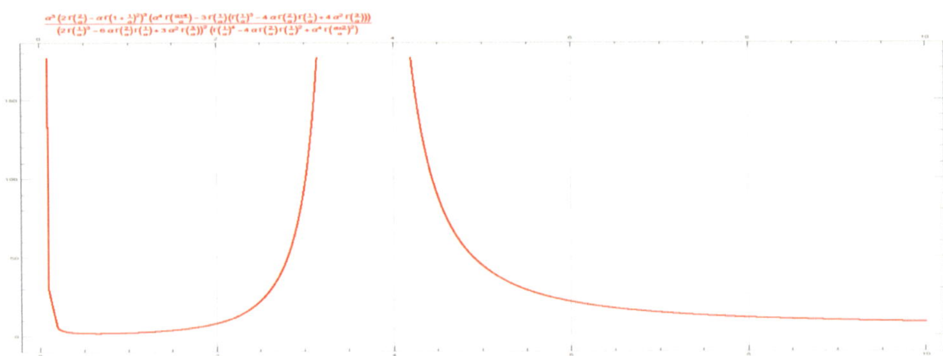

Figure 16.
Plot of Kumaraswamy distribution shape factor boundary value for given α when $\beta \to \infty$.

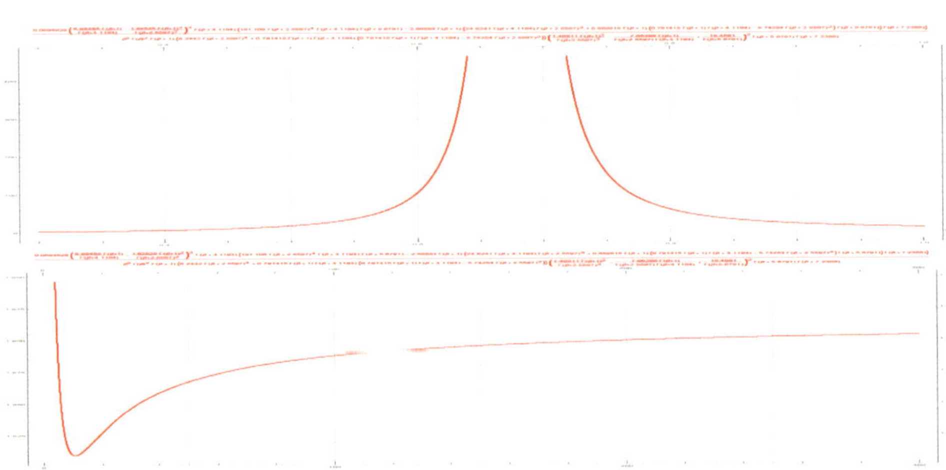

Figure 17.
Plot of Kumaraswamy distribution shape factor for given α = 0.641149, β in the range 0.3–1 and 1–300.

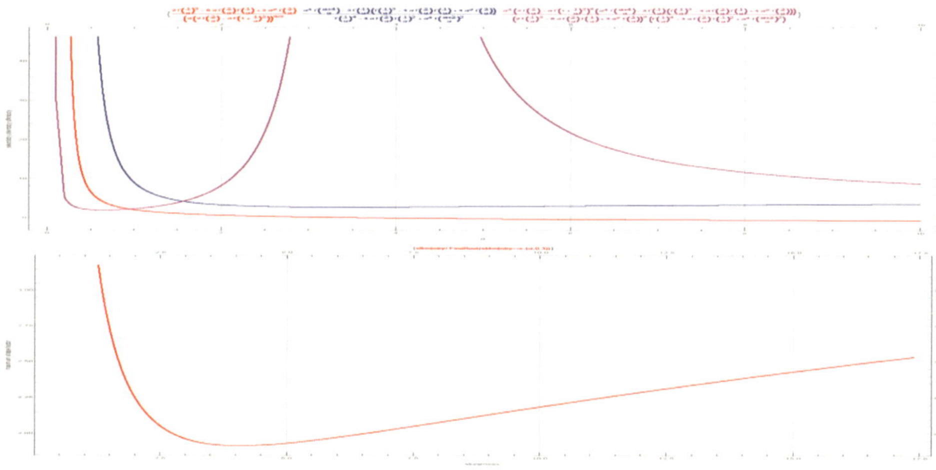

Figure 18.
Plot of Eq. (6)–(8) and plot of Kumaraswamy distribution maximum shape factor for given skewness.

$\alpha \to 0$, or at some middle point where the contour plot of the skewness and the contour plot of the kurtosis will be tangent to each other. The Mathematica contour plot does not work for a very small α, but by numerical minimization we know the global minimum of the Kumaraswamy distribution shape factor is 1.03709 when $\alpha = 1.80143 * 10^{-9}$, $\beta = 0.247044$. The conditional minimum of the shape factor when skewness = 5.99378 is about 1.04753 when $\alpha = 10^{-10.5}$, $\beta = 0.149286$ through list calculation; this is higher than 1 + 1/S^2 = 1 + 1/5.99378199789956^2 = 1.02784, the lower boundary of Beta distribution.

The Mathematica contour plot works for large α, and we see the shape factor is increasing along the contour of skewness, which attains its maximum when $\beta \to \infty$. For example, for NATH skewness 5.99378199789956, the maximum shape factor is 1.97131, arriving at α = 0.5239510562868946. The maximum shape factor of Kumaraswamy distribution for given skewness is in **Figure 18**, which is algebraically represented by the parametric curve of Eq. (6) and Eq. (8).

So the permissible shape factor range of the Kumaraswamy distribution still spans the lower end of the whole allowable range of $(1,\infty)$, but higher than that of the Beta distribution. Affine transformed Kumaraswamy distribution can fit all the first four moments of NATH, with the fitted distribution TVaR close to NATH TVaR in the error range of 5–6%, while the best effort affine transformed Beta distribution is in the error range of 9–10%.

To further improve the fit, we need additional freedom in parameters, such as the *GB*1 distribution [15], since $KumaraswamyDistribution[\alpha, \beta] \approx GeneralizedBetaDistributionI[1, \beta, \alpha, 1]$, and the maximum shape factor plot in **Figure 18** is lower than that of LogNormalDistribution, the upper bound of *GB*1. The following section will study a sibling distribution to *GB*1, fitted as good as *GB*1, but is more widely known.

5. BetaPrime distribution

Beta distribution and Kumaraswamy distribution are a few exceptions which have analytical formulas for the shape factor bounds; for other distributions to be studied, numerical optimization and empirical plot or formula will be the only feasible approach.

Transformation of Beta Distribution by x/(1-x) is the *GB*2([15]), or $BetaPrimeDistribution[p = \alpha, q = \beta, \alpha = 1, \beta = 1]$ ([11]): $TransformedDistribution\left[\frac{x}{1-x}, x \approx BetaDistribution[\alpha, \beta]\right] \approx BetaPrimeDistribution[\alpha, \beta]$. The minimum shape factor of Beta Distribution is 1, but that of the transformed is 1.5:

$$NMinimize\left[\left\{\frac{3(-3+\beta)\left(2(-1+\beta)^2+\alpha^2(5+\beta)+\alpha(-1+\beta)(5+\beta)\right)}{4(-4+\beta)(-1+2\alpha+\beta)^2}, \alpha > 0, \beta > 4\right\}, \{\alpha, \beta\}\right] = \{1.5000000239052607, \{\alpha \to 0., \beta \to 6.274769836372949 \times 10^7\}\}.$$

Empirically, the larger the third parameter α, the smaller the minimum shape factor. The smallest shape factor we get of the *BetaPrimeDistribution* is 1.125, when α = 446.49537:

$$FindMinimum\left[\left\{\frac{Kurtosis[BetaPrimeDistribution[p, q, \alpha, \beta]]}{Skewness[BetaPrimeDistribution[p, q, \alpha, \beta]]^2} /.\alpha \to \frac{1}{x} /.q \to 4\,x\right.\right.$$

$$\left.+y /.x \to 10^z, 1. > p > 0., y > 1., -4. < z < -1.\right\}, \{\{p, 6.384125235007732 \times 10^{-10}\},$$

$$\{y, 1.0032844709998097\}, \{z, -2.157370895027263\}\}, MaxIterations \to 5000]$$

$$= \{1.1250258984236121, \{p \to 2.083731454230264 \times 10^{-8},$$

$$y \to 42{:}816363091057056, z \to -2.6498169598310573\}\}.$$

This is the same value as the minimum shape factor for $GammaDistribution[\alpha, \beta, \gamma, \mu]$ (in Section 6). When α > 10,000, the Gamma function involved will not calculate or will calculate incorrectly.

With the transformation of p-> 10^w, α-> 10^-z, q-> 4*10^z + y, we can study the *GB*2 shape factor change tendency with respect to α, **Figure 19**, and shape factor change tendency with respect to p, **Figure 20**.

The *GB*2 shape factor is mainly determined by α and p, only slightly changing with respect to q when q is smaller than 5. The change with respect to α and p is similar, having two peaks, or three peaks if we regard the two sides of the infinity as

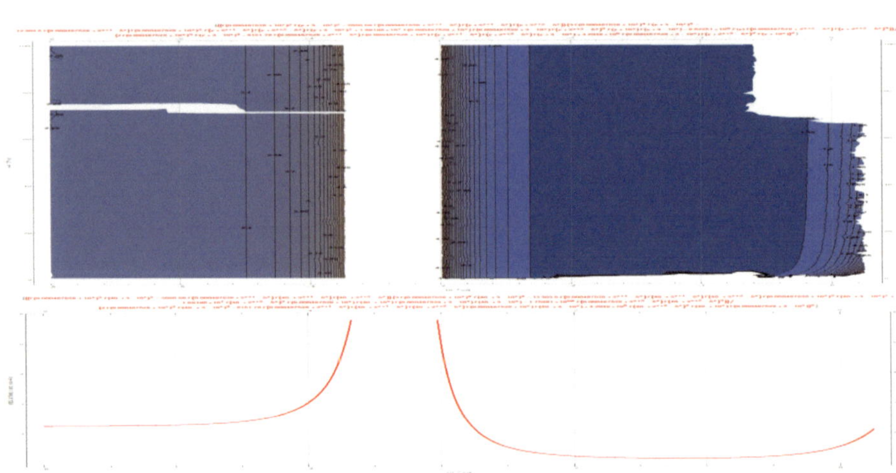

Figure 19.
GB2 distribution shape factor vs. α for fixed p = 10^-3.312 = 0.000487528.

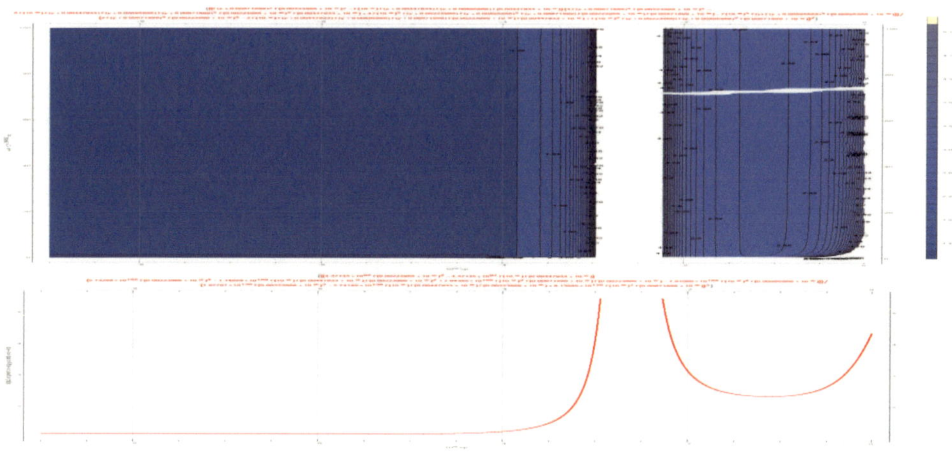

Figure 20.
GB2 distribution shape factor vs. p for fixed α = 10^2.6498169598310573 = 446.495.

two branches since that border is not easily crossable for searching or optimization algorithms.

*GB*2 shape factor's dependency with p and α, or w and z through transformation p = 10^w, α = 10^-z, is mostly unaffected by q except for right-most values of z. They are μ-shaped (**Figure 21**), this is different from Hyperbolic Distribution (in Section 8), whose shape factor dependency with λ is V-shaped. We guess V-shaped curves have unique global minimums, but μ-shaped curves will show bifurcation behavior: the converged solution in optimization will be very different when the initial point or interval is slightly different.

The knowledge that the shape factor curve attained extreme values in −3.3,-1.25 and 1 with respect to z, and attained extreme values in −2.65, −1.11 and 1 with respect to w, can be used to set the initial interval, the paramount factor determining the quality of the numerical optimization solution, for solving the *GB*2 fitting problem.

5.1 Minimum shape factor for *GB*2

The skewness and kurtosis matching problem for *GB*2 is very sensitive to the initial parameter ranges given. A study of the minimum shape factor of *GB*2 with

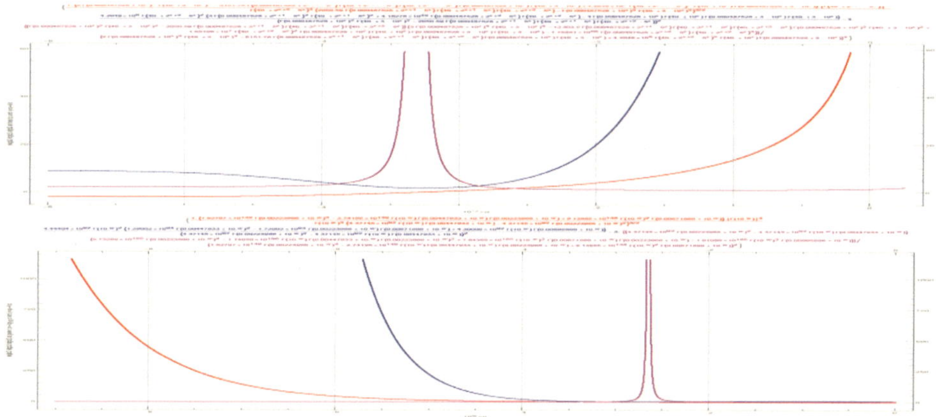

Figure 21.
GB2 distribution skewness kurtosis and shape factor vs. α or z, vs. p or w plots for fixed y = q-4/α.

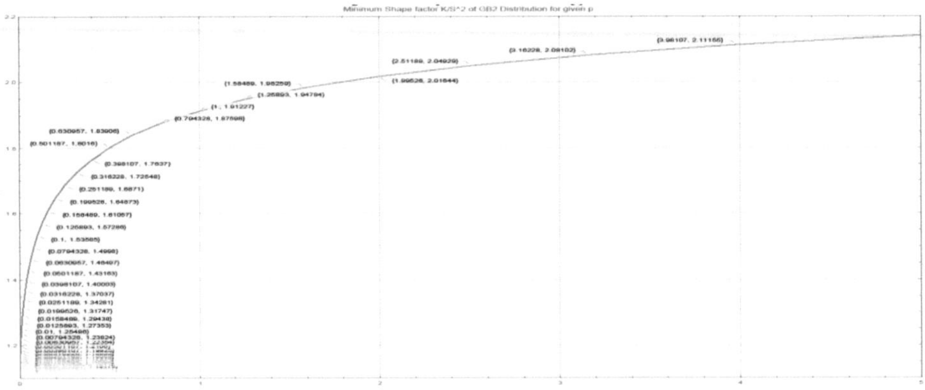

Figure 22.
The numerical minimum GB2 shape factor for given p in horizontal axis.

respect to each parameter will give us permissible ranges for those parameters. Direct work with shape factor encounters problems from Mathematica's numerical optimization function *NMinimize*, minimizing the log shape factor instead can overcome this difficulty. The plot is in **Figure 22**.

In the range (0.0001, 5.0) of p, the numerical minimum shape factor plot of *GB*2 is a very smooth curve. The fitted formula of *GB*2 min *SF* for given p by Mathematica's machine learning function *FindFormula* is Eq. (9).

$$\min \frac{K}{S^2} = 1.1593871374775397 + 1.4702458297305288 * 0.5148499158800361^{\frac{1}{p^{0.3215433282777008}}} \tag{9}$$

As a test, for NATH the log shape factor is $Log[1.83408] = 0.60654412$, the solution of Eq. (9) for p with NATH SF is $p = 0.608342$; the minimum log shape factor of *GB*2 for this p is 0.60603997, only 0.08% smaller than input.

From the contour plot **Figure 20** we know for given α, the shape factor of *GB*2 has two singular points with p or 10^w. The minimization for given α needs to carry out in each of the three regions cut by these two singular points. The plot is in **Figure 23**. With a new parameterization, $p = \frac{\lambda}{\alpha}, q = \frac{4+\nu}{\alpha}$, the minimization of shape

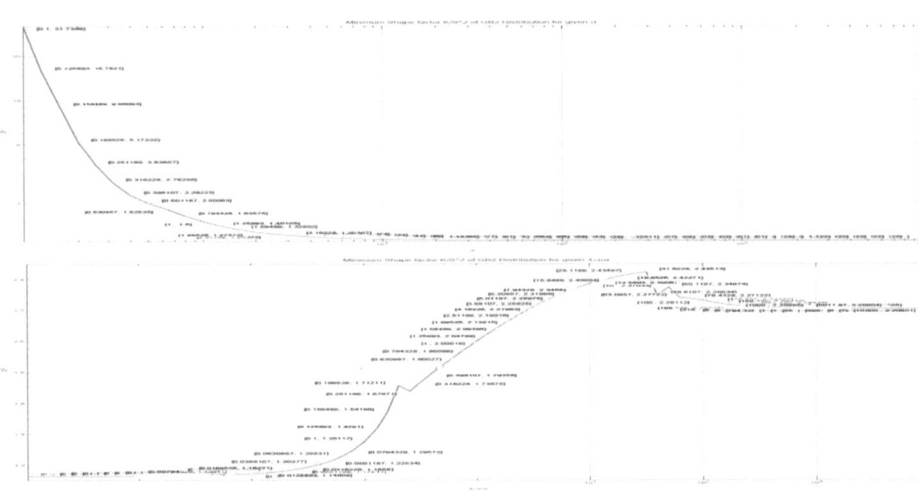

Figure 23.
The numerical minimum GB2 shape factor for given α or given pα in horizontal axis.

factor for $GB2$, for given $\lambda = p\alpha$, is easier to perform. The plot is included in **Figure 23** as well.

Figures 22 and **23** show that the permissible parameters for NATH are $p < 0.63$, $\alpha > 0.5$, $p\alpha < 0.5$. This is confirmed by $GB2$ fit practice. The best fit by $GB2$ for NATH is at $w = -0.329075005$, $p = 0.468732$, with about 5% error from input TVaR. The discontinuity of fitted $GB2$ TVaR with respect to parameter change is also observed, this w value is such a critical point.

6. Generalized gamma distribution

The generalized gamma distribution in Mathematica is the Amoroso distribution [16], with the parameter correspondence: $\alpha \leftrightarrow \alpha$, $\beta \leftrightarrow \theta$, $\gamma \leftrightarrow \beta$, $\mu \leftrightarrow a$.

For generalized gamma distribution $GammaDistribution[\alpha, \beta, \gamma, \mu]$, the shape factor depends only on α and γ. It seems the smaller the α, and the bigger the γ, the smaller the $\frac{K}{S^2}$. When $\alpha = 3.318512677036329 \times 10^{-12}$, $\gamma = 8811.572418686921$, $\frac{K}{S^2} = 1.125$, close to the global minimum 1 of K/S^2.

So there arises the question: the generalized gamma and $GB2$ can match smaller shape factors than Hyperbolic Distribution (Section 8), why they cannot fit as good as the latter for NATH with shape factor 1.83409?

One explanation is that the numerical solution for $GB2$ or generalized Gamma distribution is trapped in the shape factor curve right branch by the combined constraints of skewness and kurtosis, which is not the branch that can attain 1.125, unlike the generalized hyperbolic distribution whose shape factor has a global minimum in λ = 0.

7. Fleishman distribution

We guess 1.5 is the lower bound of shape factor for most unbounded parametric distribution families. For example, for Fleishman distribution, by the empirical formula from [5], $\gamma_4 > 1.738\gamma_3^2 - 0.3544\gamma_3 + 1.978$, the minimum shape factor is 1.72213, larger than 1.5.

The lower bound of shape factor from unbounded distributions seems, in general, to be higher than bound distributions'. Outside of the latter's upper bound and near the former's lower bound, for a *SF* value slightly larger than 1.5, in practice, most parametric distributions have difficult matching both the kurtosis and skewness: the comparatively best one is selected for study in the next section.

8. Hyperbolic distribution

Taking a sequence of numerical minimization of the shape factor, for various values of fixed λ, we get the empirical minimum shape factor curve for generalized hyperbolic distribution (GH), $HyperbolicDistribution[\lambda, \alpha, \beta, \delta, \mu]$, in **Figure 24**.

We observed that when λ > -0.6, the minimum shape factor is attained when α^2-β^2-> 0 and β-> 0, that is, it is attained by a skew hyperbolic t distribution [17–19]. When looking at the plot of shape factor with respect to λ, we feel that it must have some simple formula. So we utilize Mathematica symbolic calculation to expand the shape factor with asymptotic expansion for $BesselK[\lambda, a]$, or $K_\lambda(a)$ in [20], with respect to α^2-β^2 and then take the symbolic limit, **Figure 25**.

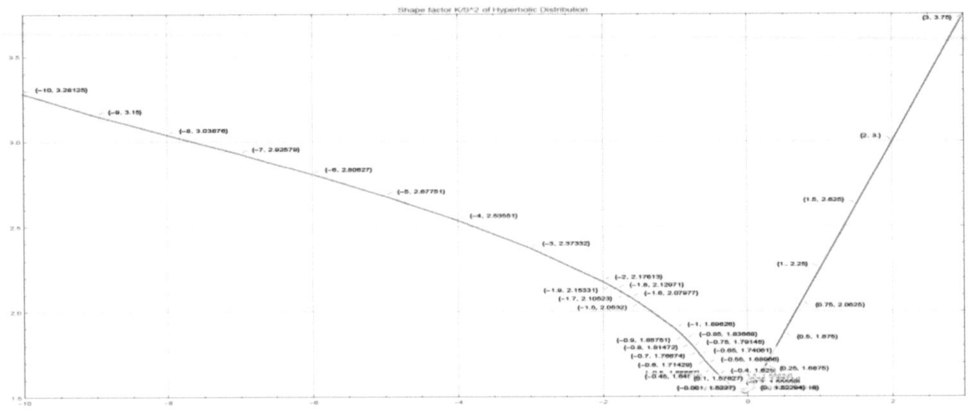

Figure 24.
V-shape of the numerical minimum GH shape factor for given λ in horizontal axis.

```
TransformedDistribution[a*u+b, u ≈ HyperbolicDistribution[λ, α, β, δ, μ]]

HyperbolicDistribution[λ, α/Abs[a], β/a, δ Abs[a], b + a μ]
```

GH is affine transformation closed, so we can assume δ=1, μ=0 for sk,kt,sf study that are affine transformation invariant.

```
sfgh = Kurtosis[#] / Skewness[#]^2 & @ HyperbolicDistribution[λ, α, β, 1, 0]

%2 //. Sqrt[α^2 - β^2] → a //. α^2 - β^2 → a^2 //. Sqrt[a^2] → a //. (a^2)^(3/2) → a //. β → b

%3 //. Sqrt[a^2] → a //. (a^2)^(3/2) → a // PowerExpand
```

When λ > 0 use Abramowitz Stegun 9.6.9 for asymptotic expansion:

```
%4 /. {BesselK[λ, a] → 1/2 Gamma[λ] (a/2)^(-λ), BesselK[1 + λ, a] → 1/2 Gamma[1 + λ] (a/2)^(-1-λ), BesselK[2 + λ, a] → 1/2 Gamma[2 + λ] (a/2)^(-2-λ), BesselK[3 + λ, a] → 1/2 Gamma[3 + λ] (a/2)^(-3-λ), BesselK[4 + λ, a] → 1/2 Gamma[4 + λ] (a/2)^(-4-λ)} // FullSimplify

Limit[%5, a → 0] // Limit[#, b → 0] & /. {Gamma[1 + λ] → λ Gamma[λ], Gamma[2 + λ] → (λ + 1) λ Gamma[λ]} // FullSimplify

%6 /. {Gamma[1 + λ] → λ Gamma[λ], Gamma[3 + λ] → (λ + 2) (λ + 1) λ Gamma[λ]} // FullSimplify

3 (2 + λ) / 4
```

When λ < 0 use Abramowitz Stegun 9.6.2 and 9.6.7 for asymptotic expansion:

```
π/2 (Gamma[-λ + 1]^-1 (a/2)^(-λ) - Gamma[λ + 1]^-1 (a/2)^λ) / Sin[λ π] /. {{λ → λ + 1}, {λ → λ + 2}, {λ → λ + 3}, {λ → λ + 4}}

{1/2 π Csc[π (1 + λ)] (2^(1+λ) a^(-1-λ)/Gamma[-λ] - 2^(-1-λ) a^(1+λ)/Gamma[2 + λ]), 1/2 π Csc[π (2 + λ)] (2^(2+λ) a^(-2-λ)/Gamma[-1 - λ] - 2^(-2-λ) a^(2+λ)/Gamma[3 + λ]), 1/2 π Csc[π (3 + λ)] (2^(3+λ) a^(-3-λ)/Gamma[-2 - λ] - 2^(-3-λ) a^(3+λ)/Gamma[4 + λ]), 1/2 π Csc[π (4 + λ)] (2^(4+λ) a^(-4-λ)/Gamma[-3 - λ] - 2^(-4-λ) a^(4+λ)/Gamma[5 + λ])}

%4 /. {BesselK[λ, a] → 1/2 π Csc[π λ] (2^λ a^(-λ)/Gamma[1 - λ] - 2^(-λ) a^λ/Gamma[1 + λ]), BesselK[1 + λ, a] → 1/2 π Csc[π (1 + λ)] (2^(1+λ) a^(-1-λ)/Gamma[-λ] - 2^(-1-λ) a^(1+λ)/Gamma[2 + λ]), BesselK[2 + λ, a] → 1/2 π Csc[π (2 + λ)] (2^(2+λ) a^(-2-λ)/Gamma[-1 - λ] - 2^(-2-λ) a^(2+λ)/Gamma[3 + λ]),
  BesselK[3 + λ, a] → 1/2 π Csc[π (3 + λ)] (2^(3+λ) a^(-3-λ)/Gamma[-2 - λ] - 2^(-3-λ) a^(3+λ)/Gamma[4 + λ]), BesselK[4 + λ, a] → 1/2 π Csc[π (4 + λ)] (2^(4+λ) a^(-4-λ)/Gamma[-3 - λ] - 2^(-4-λ) a^(4+λ)/Gamma[5 + λ])}

Limit[%9, a → 0, Direction → "FromAbove", Assumptions → -1 < λ < 0] // FullSimplify

1 + 1/(2 + λ)

Limit[%9, a → 0, Direction → "FromAbove", Assumptions → -2 < λ < -1] // FullSimplify

1 + 1/(2 + λ)
```

Figure 25.
Derivation of the GH shape factor limit when λ > −2.

The semi-empirical formula for the minimum shape factor in this range thus obtained is very simple, Eq. (10–11), which has the global minimum of 1.5 when λ turns to 0.

$$\min_{\alpha, \beta, \delta, \mu} SF = 1.5 + 0.75\lambda, \mathit{when}\ \lambda \geq 0, \tag{10}$$

$$\min_{\alpha, \beta, \delta, \mu} SF = 1 + \frac{1}{2+\lambda}, \mathit{when} - 0.6 \leq \lambda \leq 0 \tag{11}$$

When λ ≤ −0.65, however, the minimum shape factor is not attained when α^2-β^2-> 0. When λ is in the interval [−9,−0.65], the attainable smallest shape factor is between 3.15 and 1.74, with an empirical 10th order polynomial formula Eq. (12), or less accurately a mixed exponential and power function Eq. (13), found through the Mathematica *FindFormula*.

$$\begin{aligned}\min_{\alpha, \beta, \delta, \mu} SF = {} & 1.1130471668735116 - 1.6512030619809768\lambda - 1.6137376956833365\lambda^2 \\ & - 1.1485038172210114\lambda^3 - 0.5421785615853132\lambda^4 \\ & - 0.17094578834265223\lambda^5 - 0.03603744794749387\lambda^6 - 0.005000441043297472\lambda^7 \\ & - 0.0004372189547557593\lambda^8 - 0.000021791071048963054\lambda^9 \\ & - 4.711954312790356 \times 10^{-7}\lambda^{10}\end{aligned} \tag{12}$$

$$\begin{aligned}\min_{\alpha, \beta, \delta, \mu} SF = {} & 2.2104215691249425 - 0.6522131009473879*1.6355318649123258^{\lambda} \\ & + \frac{0.018965779149540653}{\lambda^3} - 0.1051542360603726\lambda\end{aligned} \tag{13}$$

So for each given K/S^2 value, there exists a permissible interval of λ, whose lower bound is calculated via Eq. (11–12) and upper bound is calculated via Eq. (10). When λ changes inside this interval, we noticed that the 0.99 TVaR of the first four moments matched generalized hyperbolic distribution will increase with respect to λ. If the lower bound still has 0.99 TVaR bigger than the input TVaR, then it is not possible to fit with moments matched *HyperbolicDistribution*. The opposite statement is also valid for the interval upper bound.

With this knowledge, the NATH permissible λ interval is [−0.8439,0.4454], and the left end point still have 0.99 TVaR larger than the input TVaR, but now only by 4.05%, better than the 5% error of *GB*2.

9. Conclusion and discussions

We proposed using the ratio of kurtosis by squared skewness as the best candidate for shape factor that can characterize the distribution asymmetry, as well as the PDF steepness. The closer this factor to 1, the more asymmetry and the steeper the PDF. The asymptotic approximation and symbolic limit is used to calculate the boundary of this factor for various distributions: the Beta, the Kumaraswamy, and the Hyperbolic Distributions, for example. This range information of the shape factor, with the surprisingly simple formulas in the three above examples (Eqs. 5–8, 10, 11), can be used to select or eliminate candidate distributions for fitting. The plot of the shape factor together with plot for skewness and kurtosis can aid in setting the initial value or parameter intervals when fitting distribution to data by numerical optimization, which usually would not work well without this information.

The idea of the shape factor and the careful study of each distribution for this shape factor is the preliminary for the numerical optimization that finally finds the best fit. The information provided by shape factor plot is rough but the numerical optimization's dependence on initial value or intervals is delicate, exemplified by *GB*2 case. The optimization function *NMinimize* and *FindMinimum* in Mathematica sometimes can only find a local optimum at best. As shown in [21, 22], the DyHF and the CMODE algorithms are the two best no-adjustment-needed global optimization algorithms. Now that the C^2oDE algorithm is better than these two [23], it would be desirable to see how it works on the GB2 fit problem. With a foolproof universally applicable global optimization algorithm, the ado with shape factor and their boundaries will no longer be needed, or be used merely as some validations; but before that time, the hard earned knowledge about shape factor through CAS is still indispensable. This is a good topic for subsequent research.

Our shape factor idea is only a small step ahead of the skewness-kurtosis plot of Pearson [6] and McDonald *et al* [15, 24–26]. Or we just made the idea implicitly in their plot explicit. But with this clearly defined form, anyone can readily start calculating it for any interested candidate distribution.

Our formula Eq. (5) is not new, since Beta distribution has the same range of skewness, kurtosis, and shape factor as the scaled Beta distribution, the *B*1 distribution in [15]. Our presentation is an example of how our method can be used to easily arrive at those formulas. Theoretically equivalent expressions are not equivalent in application. With data distributions usually not having small skewness, Eq. (5) says that the Beta distribution has a shape factor roughly in the range of (1, 1.5), this not only reveals an intrinsic property of Beta distribution, but is also more easily applicable in practice than the skewness-kurtosis plot.

The residual error of all the distributions tested so far indicates that the power function or simple exponential function PDF is not enough to provide the additional freedom of shifting for the EP curve on the condition of matched first four moments. Other forms such as mixtures, combinations, or transformations of distributions may need to be considered. A previous study indicated the following transformations are good candidates [4, 27–32]: EWGU, KGG, EG, EWED, LIG, THT. Further research will be done along these lines.

Acknowledgements

This research is supported by Validus Research Inc. The author thanks his colleagues and former colleagues for helpful suggestions and feedbacks, the editor for revision suggestions, and Claire Wang for help with grammar and language corrections.

Conflict of interest

The author declares no conflict of interest.

Author details

Frank Xuyan Wang
Validus Research Inc., Waterloo, Ontario, Canada

*Address all correspondence to: frank.wang@validusresearch.com

References

[1] Sharma K. Natural Catastrophe Modeling for Pricing in Insurance [Thesis]. Tartu: University of Tartu; 2014

[2] Currie ID. Maximum likelihood estimation and mathematica. Applied Statistics. 1995;**44**(3):379-394

[3] Wang F. Dfittool for Mathematica [Internet]. 2016. Available from: https://web.archive.org/web/20181109205350/https://www.linkedin.com/pulse/dfittool-mathematica-wang-frank/ [Accessed: 2018-11-09]

[4] Wang FX. An inequality for reinsurance contract annual loss standard deviation and its application. In: Salman A, Razzaq MGA, editors. Accounting from a Cross-Cultural Perspective. IntechOpen; 2018. pp. 73-89. DOI: 10.5772/intechopen.76265 Available from: https://www.intechopen.com/books/accounting-from-a-cross-cultural-perspective/an-inequality-for-reinsurance-contract-annual-loss-standard-deviation-and-its-application

[5] Ferenci T. The Use of Fleishman Distribution in the Empirical Investigation of Statistical Tests [Thesis]. Budapest University of Technology and Economics; 2012. Available from: http://www.medstat.hu/DiplomaFerenciTamasAlkMatMSc.pdf [Accessed: 2018-05-02]

[6] Pearson K IX. Mathematical contributions to the theory of evolution. —XIX. Second supplement to a memoir on skew variation. Published 1 January 1916. DOI: 10.1098/rsta.1916.0009. Available from: http://rsta.royalsocietypublishing.org/content/216/538-548/429.full.pdf [Accessed: 2018-04-23]

[7] Klaassen CAJ, Mokveld PJ, van Es B. Squared Skewness minus kurtosis bounded by 186/125 for unimodal distributions. Statistics & Probability Letters. 2000;**50**(2):131-135

[8] Wang F. Problem with BesselK Function [Internet]. 2018. Available from: https://web.archive.org/web/20181112185421/https://www.linkedin.com/pulse/problem-besselk-function-wang-frank/ [Accessed: 2018-11-12]

[9] Wolfram Mathematica Tutorial Collection. Constrained Optimization. Wolfram Research, Inc; 2008

[10] Loehle C. Global optimization using mathematica: A test of software tools. Mathematica in Education and Research. 2006:139-152

[11] Marichev O, Trott M. The Ultimate Univariate Probability Distribution Explorer [Internet]. 2013. Available from: http://blog.wolfram.com/2013/02/01/the-ultimate-univariate-probability-distribution-explorer/ [Accessed: 2018-06-06]

[12] Hill RJ, Frehlich RG. Probability distribution of irradiance for the onset of strong scintillation. Journal of the Optical Society of America. A. 1997; **14**(7):1530-1540. DOI: 10.1364/JOSAA.14.001530

[13] Olivero DE. Alpha-skew-normal distribution. Proyecciones Journal of Mathematics. 2010;**29**(3):224-240. Available from: https://pdfs.semanticscholar.org/a4f2/5b36ccbdb6845ae6fe4487ce88a87c97463b.pdf [Accessed: 2018-05-01]

[14] de Pascoa MAR, Ortega EMM, Cordeiro GM. The Kumaraswamy generalized gamma distribution with application in survival analysis. Statistical Methodology. 2011;**8**(5):411-433. DOI: 10.1016/j.stamet.2011.04.001 Available from: http://www.sciencedirect.com/science/article/pii/S1572312711000323

[15] McDonald JB, Sorensen J, Turley PA. Skewness and kurtosis properties of income distribution models. LIS working paper series, No. 569, 2011. Review of Income and Wealth. 2011. DOI: 10.1111/j.1475-4991.2011.00478.x Available from: https://pdfs.semanticscholar.org/eabd/0599193022dfc65ca00f28c8a071e43edc32.pdf

[16] Crooks GE. The Amoroso Distribution [Internet]. 2010. Available from: https://arxiv.org/pdf/1005.3274.pdf [Accessed 2018-05-03]

[17] Aas K, Haff IH. The generalized hyperbolic skew Student's t-distribution. Journal of Financial Econometrics. 2006;**4**(2):275-309. DOI: 10.1093/jjfinec/nbj006

[18] Hu W, Kercheval A. Risk management with generalized hyperbolic distributions. In: Locke P, editor. Proceedings of the Fourth IASTED International Conference on Financial Engineering and Applications (FEA '07). Anaheim, CA, USA: ACTA Press; 2007. pp. 19-24

[19] Scott DJ, Würtz D, Dong C, Tran TT. Moments of the generalized hyperbolic distribution. Computational Statistics. 2011;**26**(3):459-476. DOI: 10.1007/s00180-010-0219-z

[20] Abramowitz M, Stegun IA. Handbook of Mathematical Functions with Formulas, Graphs, and Mathematical Tables. New York: Dover; 1972

[21] Wang FX. Relay Optimization Method [Internet]. May 2014. Available from: http://www.optimizationonline.org/DB_FILE/2014/05/4345.pdf [Accessed 2018-07-28]

[22] Wang FX. Design index-based hedging: Bundled loss property and hybrid genetic algorithm. In: Tan Y, Shi Y, Buarque F, Gelbukh A, Das S, Engelbrecht A, editors. Advances in Swarm and Computational Intelligence. ICSI 2015. Lecture Notes in Computer Science, vol 9140. Cham: Springer; 2015. pp. 266-275. DOI: 10.1007/978-3-319-20466-6_29

[23] Wang BC, Li HX, Li JP, Wang Y. Composite differential evolution for constrained evolutionary optimization. IEEE Transactions on Systems, Man, and Cybernetics: Systems. . DOI: 10.1109/TSMC.2018.2807785

[24] Vargo E, Pasupathy R, Leemis LM. Moment-ratio diagrams for Univariate distributions. Journal of Quality Technology. 2010;**42**(3):1-11

[25] Celikoglu A, Tirnakli U. Skewness and kurtosis analysis for non-Gaussian distributions. Physica A: Statistical Mechanics and its Applications. 2018; **499**:325-334. DOI: 10.1016/j.physa.2018.02.035

[26] Huerlimann W. Normal variance-mean mixtures (I) an inequality between skewness and kurtosis. Advances in Inequalities and Applications. 2014;**2014**(2)

[27] Cordeiro GM, Ortega EMM, Ramires TG. A new generalized Weibull family of distributions: Mathematical properties and applications. Journal of Statistical Distributions and Applications. 2015;**2**(13). DOI: 10.1186/s40488-015-0036-6

[28] Barreto-Souza W, Santos AHS, Cordeiro GM. The Beta generalized exponential distribution. Journal of Statistical Computation and Simulation. 2010;**80**:159-172 https://arxiv.org/abs/0809.1889v1

[29] Lemonte AJ, Cordeiro GM. The exponentiated generalized inverse Gaussian distribution. Statistics & Probability Letters. 2011;**81**(4): 506-517. DOI: 10.1016/j.spl.2010.12.016

[30] Alzaghal A, Famoye F, Lee C. Exponentiated T-X family of distributions with some applications. International Journal of Statistics and Probability. 2013;**2**(3). DOI: 10.5539/ijsp.v2n3p31

[31] Okorie IE, Akpanta AC, Ohakwe J, Chikezie DC, Shiraishi H. The modified power function distribution. Cogent Mathematics. 2017;**4**(1). DOI: 10.1080/23311835.2017.1319592

[32] Borzadaran GR, Borzadaran HAM. Log-concavity property for some well-known distributions. Surveys in Mathematics and its Applications. 2011; **6**:203-219

Chapter 4

Topological Properties and Dynamic Programming Approach for Designing the Access Network

Franco Robledo, Pablo Romero, Pablo Sartor, Luis Stábile and Omar Viera

Abstract

A wide area network (WAN) can be considered as a set of sites and a set of communication lines that interconnect the sites. Topologically a WAN is organized in two levels: the backbone network and the access network composed of a certain number of local access networks. Each local access network usually has a treelike structure, rooted at a single site of the backbone and connected users (terminal sites) either directly to this backbone site or to a hierarchy of intermediate concentrator sites which are connected to the backbone site. The backbone network has usually a meshed topology, and this purpose is to allow efficient and reliable communication between the switch sites that act as connection points for the local access networks. This work tackled the problem of designing the Access Network Design Problem (ANDP). Only the construction costs, e.g., the costs of digging trenches and placing a fiber cable into service, are considered here. Different results related to the topological structure of the ANDP solutions are studied. Given the complexity of the ANDP (the problem belongs to the NP-hard class), recurrences to solve it are proposed which are based on Dynamic Programming and Dynamic Programming with State-Space Relaxation methodology.

Keywords: topological design, access network, dynamic programming with state-space relaxation

1. Introduction

Telecommunication networks have become strategic resources for private- and state-owned institutions, and its economic importance continuously increases. There are series of recent tendencies that have a considerable impact on the economy evolution such as growing integration of networks in the productive system, integration of different services in the same communication system, and important modification in the telephone network structure. Such evolutions accompany a significant growth of the design complexity of these systems. The integration of different sorts of traffics and services and the necessity of a more accurate management of the service quality are factors that make this type of systems very hard to design, to dimension, and therefore to optimize. This situation is aggravated with a very high competitiveness context in an area of critical strategic importance.

The conception of a WAN is a process in which dozens of sites with different characteristics require to be connected in order to satisfy certain reliability and performance restrictions with minimal costs. This design process involves the terminal site location, the concentrator location, the backbone (central network or kernel) design, the routing procedures, as well as the lines and nodes dimensioning. A key aspect on WAN design is the high complexity of the problem, as much in its globality as in the principal subproblems in which it is necessary to decompose it. Due to the high investment levels, a cost decrease of very few percentage points while preserving the service quality results in high economic benefits.

Typically, a WAN network global topology can be decomposed into two main components: the access network and the backbone network. These components have very different properties, and consequently they introduce specific design problems (although they are strongly interdependent). On the one hand, this causes complicated problems (particularly algorithmic ones); on the other hand, it leads to stimulating and difficult research problems.

A WAN access network is composed of a certain number of access subnetworks, having treelike topologies; and the flow concentration nodes allow to diminish the costs. These integrated flows reach the backbone which has a meshed topology, in order to satisfy security, reliability, vulnerability, survivability, and performance criteria. Consequently, the backbone is usually formed by high-capacity communication lines such as optic fiber links.

Modeling a WAN design by means of the formulation of a single mathematical optimization problem is very intricate due to the interdependence of its large amount of parameters. Therefore the design of a WAN is usually divided into different subproblems [1–4]. A good example of a possible decomposition approach for the WAN design process is the following [5]:

1. Access and backbone network topologies design. Specific knowledge about the cost of laying lines between different network sites (terminals, concentrators, and backbone) is assumed. Frequently, these costs are independent of the type of line that will effectively be installed since they model the fixed one-time costs (cost of digging trenches in the case of optic fiber, installing cost, placing a fiber cable into service). A high percentage from the total construction network budget is spent in this phase [6].

2. Dimensioning of the lines that will connect the different sites of the access and backbone networks and the equipment to be settled in the mentioned sites.

3. Definition of the routing strategy of the flow on the backbone network.

This work focuses on phase (1) of the decomposition of a WAN design process. More precisely, it deals with the topology planning process concerning the access network. Due to the NP-hard nature of the problem and even though there exist some results, there is still room for improving industrial practices applied today. In this sense, the authors believe it is of strategic importance to design powerful quantitative analysis techniques, potentially easy to integrate into tools. Combinatorial optimization models are introduced that formally define the topological design of the access networks. Moreover, different results related to the topological structure are introduced. Finally, different algorithms are proposed for the topological design which are based on Dynamic Programming and Dynamic Programming with State-Space Relaxation methodology.

2. A model for a WAN design

In this section, a model for the design of a WAN is introduced. The model tries to show the most essential aspects which are considered when designing access and backbone networks. In this model, some parameters are not considered: the operation probability of the lines and equipment, the number of equipment ports, and the memory capacity of the equipment. The objective is to design a WAN with the smallest possible installation cost, so that the constraints are satisfied.

In what follows, the data of the model are presented as well as its formalization as a combinatorial optimization problem on weighted graphs. The goal is to find the optimal topology that satisfies the imposed constraints to the access and backbone networks. **Figure 1** shows an example of a wide area network. The information available for each type of equipment (switch and concentrator) and each type of connection line, as well as the line laying, is the following:

- E_a is the set of types of connection lines available. Furthermore $\forall e \in E_a$ the following data are given:
  - c_e is the cost by kilometer of the line type e. Here the laying cost is not included.
  - v_e is the speed in Kbits/s of the line type e.
- K is the set of types of concentrator equipment available. Furthermore $\forall k \in K$ the following data are given:

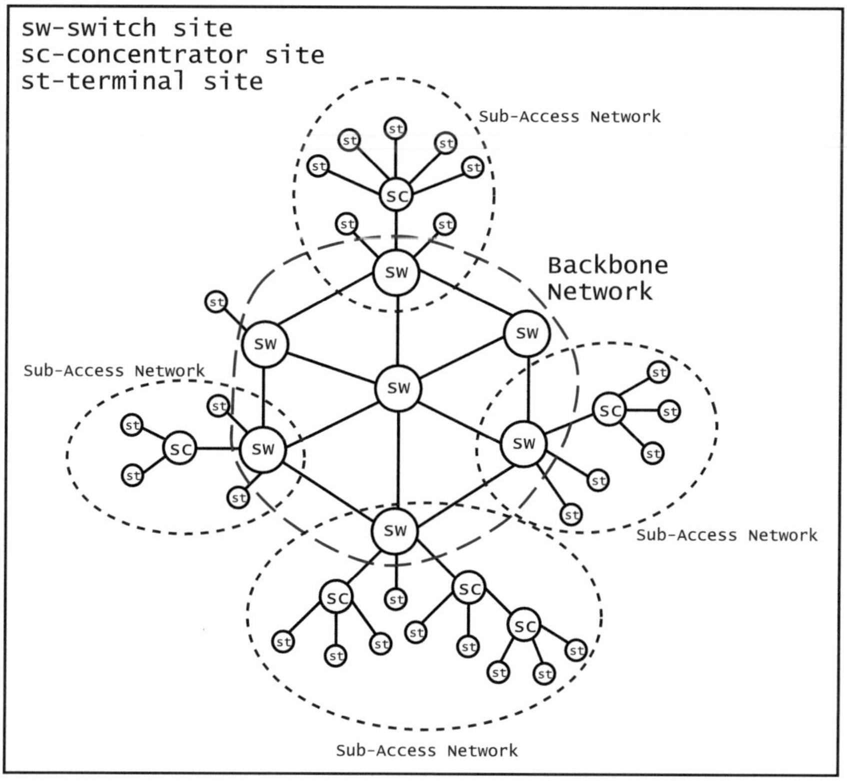

Figure 1.
WAN example.

- c_k is the installation cost of the concentrator type k.
- v_k is the speed in Kbits/s of the concentrator type k.

- W is the set of types of switch equipment available. Furthermore $\forall w \in W$ the following data are given:
  - c_w is the installation cost of the switcher type w.
  - v_w is the speed in Kbits/s of the switcher type w.

- $C = F_{cost}(L) = \{c_{ij} = direct\ connection\ costs\ between\ the\ sites\ i,j; \forall i \in S, \forall j \in S_C \cup S_D\}$; this matrix gives us, for a site of S and a site of $S_C \cup S_D$, the cost of laying a line among them. When the direct connection among both places is not possible, we assume that $c_{ij} = \infty$.

In terms of graph theory, a model for the design of a WAN, based on the problem, is presented as follows. Some notation is introduced next, that is then used to formally define the problem.

- $E_1 = \{(i,j); \forall i \in S_T, \forall j \in S_C \cup S_D / d_{ij} < \infty\}$ is the set of feasible connections between a terminal site and a concentrator or switch site.
- $E_2 = \{(i,j); \forall i \in S_C, \forall j \in S_C \cup S_D / d_{ij} < \infty\}$ is the set of feasible connections between a concentrator site and a switcher or another concentrator site.
- $E_3 = \{(i,j); \forall i \in S_D, \forall j \in S_D / d_{ij} < \infty\}$ is the set of feasible connections between two switch sites.
- $E = E_1 \cup E_2 \cup E_3$ is the set of all feasible connections on the WAN.
- $D_{S_T} = \{D_{t_i}, t_i \in S_T\}$, where • D_{t_i} is the set of terminal nodes which demand connections with $t_i \in S_T$.
- $V_{S_T} = \{v_{i,j}\}_{i,j \in S_T}$ is the traffic demand matrix.

Definition 1 (WANDP—wide area network design problem). Let $G = (S,E)$ be the graph of feasible connections on the WAN. The wide area network design problem $(S,E,K,W,E_a,C,D_{S_T},V_{S_T})$ consists in finding a subnetwork of G of minimum cost which satisfies the following points:

1. The backbone network topology must be at least 2-node-connected.
2. The access and backbone networks must be able to support the demand of connection and traffic required by the terminal sites.

Given the complexity of the WANDP, to facilitate its solution, the topological design problem is divided into three subproblems:

1. The Access Network Design Problem
2. The backbone network design problem (BNDP)

3. The routing (or flow assignment) and capacity assignment problem (RCAP)

The remainder of this work concentrates only in the first problem (ANDP).

3. Access Network Design Problem

The Access Network Design Problem is defined as follows.

Definition 2 (ANDP—Access Network Design Problem). Let $G_A = (S, E_1 \cup E_2)$ be the graph of feasible connections on the access network and C the matrix of connection costs defined previously. The Access Network Design Problem $(S, E_1 \cup E_2, C)$ consists in finding a subgraph of G_A of minimum cost such that $\forall i \in S_T$; there exists a path from i to some site $j \in S_D$ of the backbone network.

Notation 1. Γ_{ANDP} denotes the space of feasible solutions of ANDP$(S, E_1 \cup E_2, C)$ that do not have any cycle and with an output only toward the backbone network $\forall t \in S_T$. These have forest topology as we illustrate in **Figure 2**.

In order to define these problems in terms of graph theory, the following notation is introduced:

- S_T is the set of terminal sites (clients) to be connected to the backbone.
- S_C is the set of feasible concentrator sites of the access network. On each one of these sites, an intermediate server equipment might be placed. From this one, a trunk line is laid toward the backbone or other concentrator site.
- S_D is the set of feasible switch sites of the backbone network. On each one of these sites, a powerful server might be placed and, from it, connection lines toward other backbone server equipment.
- $V = S_T \cup S_C \cup S_D$ are all the feasible sites of the WAN network.
- $A = \{a_{ij}\}_{i,j \in V}$ is a matrix which gives for any pair of sites $i, j \in V$, the cost $a_{ij} \geq 0$ of laying a line between them. When the direct connection between i and j is not possible, we define $a_{ij} = \infty$.

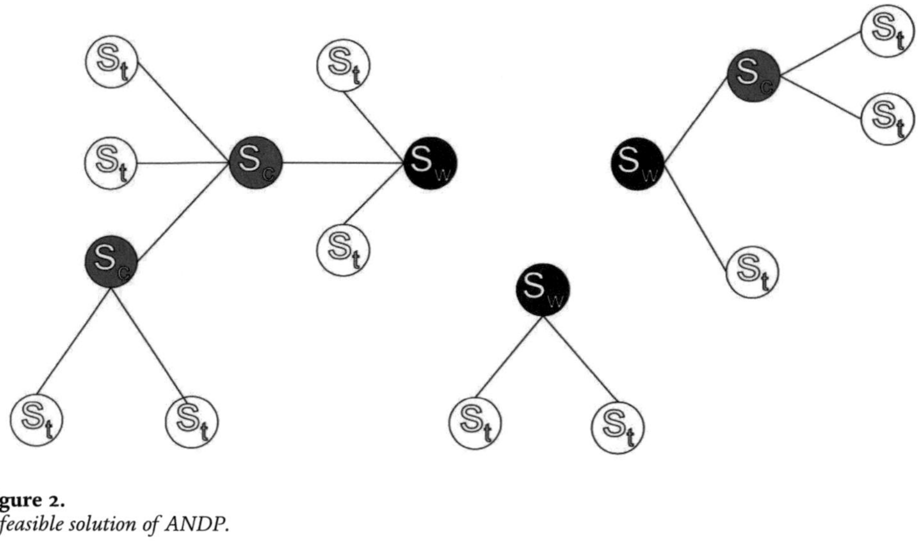

Figure 2.
A feasible solution of ANDP.

- $U = \{(i,j)|i,j \in V, a_{ij} < \infty\}$ is the set of all the feasible connections between the different sites of the WAN network.

- $G = (V, U)$ is the simple graph which models every node and feasible connection of the WAN.

The General Access Network Design Problem (GANDP) consists of finding a minimum-cost subgraph $H \subset G$ such that all the sites of S_T are communicated with some node of the backbone. This connection can be direct or through intermediate concentrators. The use of terminal sites as intermediate nodes is not allowed; this implies that they must have degree one in the solution.

The GANDP is here simplified by collapsing the backbone into a fictitious node and given the name of "Access Network Design Problem." The equivalence between both problems, GANDP and ANDP, as well as the NP-hardness of the ANDP, is proved in [7].

This work concentrates on the ANDP with the objective of proposing a new approach for solving this problem. We study different results related to the topological structure of the ANDP solutions. In particular we present results that characterize the topologies of the feasible solutions of an ANDP instance. The following proposition shows the topological form of the feasible solutions of ΓANDP for a given ANDP instance.

Proposition 1. Given an ANDP with associated graph $G_A = (S, E_1 \cup E_2)$ and matrix of connection costs C. If the subnetwork $H = (S_T \cup \overline{S}, \overline{E})$ (with $\overline{S} \subseteq S_C \cup S_D$ and $\overline{E} \subseteq E_1 \cup E_2$) is an optimal solution of ΓANDP, it is composed of a set of disjoint trees $H = \{H_1, \ldots, H_m\}$ that satisfy:

1. $\forall H_l \in H, \exists j \in S_D$ *unique* $/ j \in H_l$

2. $\forall H_l \in H, \exists$ *a subset* $S_T^l \subset S_T, S_T^l \neq \varnothing_{\overline{S_T^l}} \subseteq NODES(H_l)$

3. $\bigcup_{l=1}^{m} S_T^l = S_T$

Proof. Trivial.

The following propositions present results that characterize the structure of the global optimal solution.

Proposition 2. Let ANDP $(S, E_1 \cup E_2, C)$ be a problem where $s_c \in S_C, \overline{s} \in S_C \cup S_D$ and $s \in S_T \cup S_C$ such that $\{(s, s_c), (s_c, \overline{s})\} \subset E_1 \cup E_2$ and $\exists s_w \in S_D / c_{s,s_w} < c_{s,s_c} + c_{s_c,\overline{s}}$. Then, if $T_A \in$ ΓANDP is a globally optimal solution, it is fulfilled that $g(s_c) \geq 3$ in T_A, $\forall s_c \in T_A, s_c \in S_C$.

Proof. Let us suppose that there exists $T_A \in$ ΓANDP global optimal solution such that $\exists s_c \in T_A$ a concentrator site with $g_{s_c} < 3$ in T_A. If $g(s_c) = 1$; then s_c is a pendant in T_A; therefore, eliminating this, a feasible solution of smaller cost would be obtained. This is a contradiction; hence, $g(s_c) \neq 1$. If $g(s_c) = 2$, let $\overline{s} \in S_C \cup S_D$ be the site adjacent to s_c in T_A which its output site is toward the backbone network. Let $s \in S_T \cup S_C$ be the other adjacent site in T_A. Considering the network $H = (T_A[s_c]) \cup \{(s, s_w)\}$, where $s_w \in S_D$ satisfies $c_{s,s_w} < c_{s_c,\overline{s}}$, it is fulfilled:

$$COST(H) = COST(T_A) - c_{s,s_c} - c_{s_c,\overline{s}} + c_{s,s_w} < COST(T_A) \quad (1)$$

Furthermore, it is easy to see that $H \in$ ΓANDP. Hence, this implies that H is a better feasible solution compared with T_A. This is a contradiction, entailing that $g(s_c) \geq 3$ in T_A, as required and completing the proof.

Proposition 3. Given an ANDP $(S, E_1 \cup E_2, C)$ such that for any three sites (s_1, s_2, s_3), with $s_1 \in S_T \cup S_C$, $s_2 \in S_C$ and $s_3 \in S_C \cup S_D$, the strict triangular inequality is satisfied, i.e., $c_{s_1,s_k} < c_{s_i,S_j} + c_{s_j,s_k}$, $i, j, k \in \{1, 2, 3\}$. Then, if $T_A \in \Gamma_{\text{ANDP}}$ is a globally optimal solution, it is fulfilled that $g(s_c) \geq 3$ in T_A, $\forall s_c \in T_A$, $s_c \in S_C$.

Proof. As in the previous proposition, let us suppose that there exists $T_A \in \Gamma_{\text{ANDP}}$ global optimal solution such that $\exists s_c \in T_A$, a concentrator site with $g(s_c) < 3$ in T_A. Clearly $g(s_s)$ must be different to 1. Now, let us consider the case $g(s_c) = 2$ in T_A. Let s_1, s_2 be the adjacent sites to s_c in T_A. By hypothesis $c_{s_1,s_2} < c_{s_1,s_c} + c_{s_c,s_2}$. Considering the network $\overline{T_A} = (T_A\{s_c\}) \cup \{(s_1, s_2)\}$, a feasible solution is found, and moreover

$$COST(\overline{T}_A) = COST(T_A) - c_{s_1,s_c} - c_{s_c,s_2} + c_{s_1,s_2} < COST(T_A) \tag{2}$$

This is a contradiction; therefore, $g(s_c) \geq 3$ in T_A, hence completing the proof.

The next section presents algorithms applied to the ANDP$^{(\leq k)}$ with $k \in \{1, 2\}$. A way of computing the global optimal solution cost of it using the Dynamic Programming approach is obtained. Considering that the ANDP$^{(\leq 1)}$ is a NP-hard problem, we obtain lower bounds to the global optimal solution cost by Dynamic Programming with State-Space Relaxation in polynomial time.

4. Algorithms applied to the ANDP

This chapter presents the Dynamic Programming approach as alternative methodology to find a global optimal solution cost for the ANDP$^{(\leq 1)}$ and ANDP$^{(\leq 2)}$. After we introduce the Dynamic Programming with State-Space Relaxation as a method to obtain lower bounds for the original problem.

4.1 Dynamic Programming

Proposition 4. Given an ANDP $(S, E_1 \cup E_2, A)$, the cost of a global optimal solution of $\Gamma^{\leq 1}_{ANDP}$ is given by $f_{(S_T, Z, A^Q)}$, with $f_{(.,.,.)}$ defined by the following expression of Dynamic Programming:

$$f_{S_C}(S_T, Z, A^Q) = \begin{cases} \min\limits_{s_t \in S_T} \left\{ \begin{array}{c} COST(s_t, Z) + f_{S_C}(S_T\{s_t\}, Z, A^Q), \\ \min\limits_{s_c \in S_C} \left\{ \begin{array}{c} COST(s_t, s_c) + COST(s_c, Z) + \\ f_{S_C}\left(S_T\{s_t\}, Z, A^{Q \cup \{(s_c, Z)\}}\right) \end{array} \right\} \end{array} \right\} & if\, S_T \neq \varnothing \\ 0 & otherwise \end{cases} \tag{3}$$

where $COST(s, Z) = \min_{z \in S_D} \{COST(s, z)\}$, $(s, Z) = \text{argmin}_{z \in S_D} \{COST(s, z)\}$ and the matrix of connection costs $A^Q = \{a_{i,j}\}_{i,j \in E_1 \cup E_2}$ is defined by

$$a_{i,j} = \begin{cases} COST(i,j) & if\, (i,j) \notin Q \\ 0 & otherwise \end{cases} \tag{4}$$

Proposition 5. Given an ANDP $(S, E_1 \cup E_2, A)$, the cost of a global optimal solution of $\Gamma^{\leq 2}_{ANDP}$ is given by $f_{(S_T, Z, A^Q)}$, with $f_{(.,.,.)}$ defined by the following expression of Dynamic Programming

$$f_{S_C}(S_T, Z, A^Q) = \begin{cases} \min_{s_t \in S_C} \left\{ \begin{array}{c} COST(s_t, Z) + f_{S_C}(S_T\{s_t\}, Z, A^Q), \\ \min_{s_c \in S_C} \left\{ \begin{array}{c} COST(s_t, s_c) + COST(s_c, Z) + \\ f_{S_C}\left(S_T\{s_t\}, Z, A^{Q \cup \{(s_c, Z)\}}\right) \end{array} \right\}, \\ \min_{(s_c^u, s_c^v) \in E_2} \left\{ \begin{array}{c} COST(s_t, s_c^u) + \\ COST(s_c^u, s_c^v) + COST(s_c^v, Z) + \\ f_{S_C}\left(S_T\{s_t\}, Z, A^{Q \cup \{(s_c^u, s_c^v), (s_c^v, Z)\}}\right) \end{array} \right\} \end{array} \right\} & if S_T \neq \varnothing \\ 0 & otherwise \end{cases} \quad (5)$$

where $COST(s, Z) = \min_{z \in S_D} \{COST(s, z)\}$, $(s, Z) = \text{argmin}_{z \in S_D} \{COST(s, z)\}$ and the matrix of connection costs $A^Q = \{a_{i,j}\}_{i,j \in E_1 \cup E_2}$ is defined by

$$a_{i,j} = \begin{cases} COST(i,j) & if\,(i,j) \notin Q \\ 0 & otherwise \end{cases} \quad (6)$$

4.2 Dynamic programming with state-space relaxation

In order to find a lower bound of $f_{S_C}(S_T, Z, A^Q)$, the Dynamic Programming with State-Space Relaxation is now applied. It is a general relaxation procedure applied to a number of routing problems [8]. The motivation for this methodology stems from the fact that very few combinatorial optimization problems can be solved by Dynamic Programming alone due to the dimensionality of their state-space. To overcome this difficulty, the number of states is reduced by mapping the state-space associated with a given Dynamic Programming recursion to a smaller cardinality space. This mapping, denoted by g, must associate to every transition from a state S_1 to a state S_2 in the original state-space, a transition $g(S_1)$ to $g(S_2)$ in the new state-space. To be effective, the function g must give rise to a transformed recursion over the relaxed state-space which can be computed in polynomial time. Furthermore, this relaxation must generate a good lower bound for the original problem.

With the aim of illustrating this methodology, we present this approach in the context of the minimization of the total schedule time for the Traveling Salesman Problem with Time Window (TSPTW), after we apply it to the Dynamic Programming recursion presented in Proposition 5. The objective of the TSPTW is to find an optimal tour where a single vehicle is required to visit each of a given set of locations (customers) exactly once and then return to its starting location. The vehicle must visit each location within a specified time window, defined by an earliest service start time and latest service start time. If the vehicle arrives at a service location before the earliest service start time, it is permitted to wait until the earliest service start time is reached. The vehicle conducts its service for a known period of time and immediately departs for the location of the next scheduled customer. Assume that the time constrained path starts at fixed time value a_o. Define $F(S, i)$ as the shortest time it takes for a feasible path starting at node o, passing through every node of $S \subseteq N$ exactly once, to end at node $\imath \in S$. Note that optimization of the total arc cost would involve an additional dimension to account for the arrival time at a node. The function $F(S, i)$ can be computed by solving the following recurrence equations:

$$F(S,j) = \min_{(i,j) \in E} \{F(S - \{j\}, i) + t_{ij} | i \in S - \{j\}\} \forall S \subseteq N, j \in S \quad (7)$$

The recursion formula is initialized by

$$F(\{j\},j) = \begin{cases} \max\{a_j, a_o + t_{oj}\} & if\,(o,j) \in E \\ +\infty & otherwise \end{cases} \tag{8}$$

The optimal solution to the TSPTW is given by

$$\min_{j \in N} \{F(N,j) + t_{jd}\} \tag{9}$$

Note that Eq. (7) is valid if $a_j \le F(S,j) \le b_j$. If however $F(S,j) < a_j$, then $F(S,j) = a_j$; if $F(S,j) > b_j$, $F(S,j) = \infty$. Equations (7) and (9) define a shortest path algorithm on a state graph whose nodes are the states (S,i) and whose arcs represent transitions from one state to another. This algorithm is a forward Dynamic Programming algorithm where at step s, with $s = 1, \ldots, n+1$, a path of length s is generated. The state (S,i) of cost $F(S,i)$ are defined as follows: S is an unordered set of visited nodes and i is the last visited node, $i \in S$.

Several alternatives for the mapping g have been suggested [9]. Here is presented the shortest r-path relaxation, i.e., $g(S) = r = \sum_{i \in S} r_i$, where $r_i \ge 1$ is an integer associated with node $i \in N$; then $g(S\{i\}) = g(S) - r_i$. Define $R = \sum_{i \in S} r_i$. Hence the transformed recursion equations are

$$F(r,j) = \min_{(i,j) \in E} \{F(r - r_j, i) + t_{ij} | r - r_j \ge r_i\}, r \in \{1, \ldots, R\}, j \in N \tag{10}$$

Recursion (10) holds if $a_j \le F(r,j) \le b_j$. Otherwise, if $F(r,j) < a_j$, then $F(r,j) = a_j$; if $F(r,j) > b_j$, $F(r,j) = \infty$. The recursion formula is initialized by

$$F(\{j\},j) = \begin{cases} \max\{a_j, a_o + t_{oj}\} & if\;(o,j) \in E\; and\; q = q_j \\ +\text{\infty} & otherwise, for\; q \in \{1, \ldots, Q\}, j \in N \end{cases} \tag{11}$$

The lower bound is given by

$$\min_{j \in N} \{F(R,j) + t_{jd}\} \tag{12}$$

The complexity of the bounding procedure is $O(n^2 \times Q)$ for a n-node problem. Now, we present this approach in the context of finding a "good" lower bound for the solution of ANDP$^{(\le 2)}$. The following proposition gives a lower bound for the $f_{S_C}(S_T, Z, A^Q)$ presented in Proposition 5 (the optimum value of the ANDP$^{(\le 2)}$).

Proposition 6. Given an ANDP $(S, E_1 \cup E_2, C)$, a lower bound of $f_{S_C}(S_T, Z, A^Q)$ is derived from the following expression of Dynamic Programming with State-Space Relaxation

$$g_{S_C}(r, Z, A^Q) = \begin{cases} \min\limits_{s_t^i \in S_T} \left\{ \begin{array}{c} COST(s_t^i, Z) + g_{S_C}(r - r_i, Z, A^Q), \\ \min\limits_{s_c^j \in S_C} \left\{ \begin{array}{c} COST\left(s_t^i, s_c^j\right) + COST\left(s_c^j, Z\right) + \\ g_{S_C}\left(r - r_i, Z, A^{Q \cup \{(s_c^j, Z)\}}\right) | r - \hat{R} - r_i \ge r_j \end{array} \right\}, \\ \min\limits_{\left(s_c^{ju}, s_c^k\right) \in E_2} \left\{ \begin{array}{c} COST\left(s_t^i, s_c^j\right) + COST\left(s_c^j, s_c^k\right) + COST(s_c^k, Z) + \\ g_{S_C}\left(r - r_i, Z, A^{Q \cup \{(s_c^j, s_c^k), (s_c^k, Z)\}}\right) | \\ r - \hat{R} - r_i \ge r_j + r_k \end{array} \right\} \end{array} \right\} & if\; S_T \ne \varnothing \\ 0 & otherwise \end{cases} \tag{13}$$

where $1 \leq r_i \leq R$ is an integer associated with the site $i \in S_T \cup S_C$, $R = \sum_{i \in S_T \cup S_C} r_i$, $\hat{R} = \sum_{j \in S_C} r_j$ and the matrix of connection costs $A^Q = \{a_{i,j}\}_{i,j \in E_1 \cup E_2}$ is defined by

$$a_{i,j} = \begin{cases} COST(i,j) & \text{if } (i,j) \notin Q \\ 0 & \text{otherwise} \end{cases} \quad (14)$$

The lower bound is given by $g_{(R,Z,A^{\emptyset})}$.

5. Computational results

This section presents the experimental results obtained with the recursions of above. The algorithms were implemented in ANSI C. The experimental results were obtained in an Intel Core i7, 2.4 GHz, and 8 GB of RAM running under a home PC. The recursions presented in Propositions 4 and 5 were applied to the ANDP$^{(\leq 1)}$ and the ANDP$^{(\leq 2)}$, respectively, whereas the recursion presented in Proposition 6 was applied to ANDP$^{(\leq 2)}$. They were tested using a large test set, by modifying the Steiner Problem in Graphs (SPG) instances from SteinLib [10]. This library contains many problem classes of widely different graph topologies. Most of the problems were extracted from these classes: C, MC, X, PUC, I080, I160, P6E, P6Z, and WRP3. The SPG problems were customized, transforming them into ANDP instances by means of the following changes. For each considered problem:

1. The terminal node with greatest degree was chosen as the z node (modeling the back- bone).

2. The Steiner nodes model the concentrator sites, and the terminal nodes model the terminal sites.

3. All the edges between terminal sites were deleted (as they are not allowed in feasible ANDP solutions).

Moreover, if the resulting topology was unconnected, the problem instance was discarded. Let us notice that since in the ANDP the terminals cannot be used as intermediate nodes (which implies also that edges between pairs of terminals are not allowed), the cost of a SPG optimum is a lower bound for the optimum of the corresponding ANDP. Therefore they are for ANDP$^{(\leq k)}$ with $k \in 1...2$.

Table 1 shows the results obtained by applying the recurrences presented in Propositions 4 and 5. In each one of them, the first column contains the names of the original SteinLib classes with the name of the customized instance. The entries from left to right are:

- The size of the selected instance in terms of number of nodes, edges, and terminal sites, respectively

- A lower bound for the optimal cost; the SPG optimum cost (LB_{SPG})

- c_{opt}^1 and c_{opt}^2 where c_{opt}^k is the cost of the best feasible solution found in $\Gamma_{ANDP}^{(\leq k)}$

- The gap of the cost for the best feasible solution of $\Gamma_{ANDP}^{(\leq k)}$ with respect to the lower bound $LB_{SPG}^{(k)}$ with $k \in \{1,2\}$ $\left(LB\_GAP_{SPG}^{(k)}\right)$

Set	Name	|V|	|E|	|T|	LB_{SPG}	c_{opt^1}	c_{opt^2}	$LBGAP_{SPG^{(1)}}$	$LBGAP_{SPG^{(2)}}$
I080	i080-001	80	120	6	1787	∞	2187		22.38%
I080	i080-011	80	350	6	1479	∞	1499		1.35%
I080	i080-012	80	350	6	1484	∞	1497		0.88%
I080	i080-013	80	350	6	1381	∞	1383		0.14%
I080	i080-014	80	350	6	1397	∞	1505		7.73%
I080	i080-111	80	350	8	2051	∞	2159		5.27%
I080	i080-112	80	350	8	1885	2201	1887	16.76%	0.11%
I080	i080-113	80	350	8	1884	∞	1884		0%
I080	i080-114	80	350	8	1895	∞	2099		10.77%
I080	i080-115	80	350	8	1868	2174	1969	16.38%	5.41%
I080	i080-233	80	160	16	4354	∞	4564		4.82%
I160	i160-011	160	812	7	1677	∞	1875		11.81%
I160	i160-012	160	812	7	1750	∞	1891		8.06%
I160	i160-013	160	812	7	1661	∞	1862		12.10%
I160	i160-014	160	812	7	1778	∞	1991		11.98%
I160	i160-015	160	812	7	1768	2281	1864	29.02%	5.43%
PUC	cc3-4p	64	288	8	2338	∞	2553		9.20%
PUC	cc3-4u	64	288	8	23	∞	25		8.70%
Average								20.72%	7.01%

Table 1.
Results obtained by applying Dynamic Programming to c^1_{opt} and c^2_{opt}.

The $LB\_GAP^{(k)}_{SPG}$ is computed as

$$LB\_GAP^{(k)}_{SPG} = 100 \times \frac{c^k_{opt} - LB_{SPG}}{LB_{SPG}}. \tag{15}$$

Feasible solutions were obtained here only for i080-112, i080-115, and i160-015 with $k = 1$ because, as can be seen, the cost is finite. The optimal values of the SPG instances (LBSP G) provided lower bounds for the optimal values of the ANDP (therefore to ANDP(≤k) with $k \geq 0$), considering that in the ANDP generation process, all the connections between terminal nodes were deleted and further that ANDP's feasible solution space is more restrictive than of SPG. The experimental results obtained for c^1_{opt} have an average gap with respect to the lower bound of 20.72%. Increasing k to 2 (applying the recursion presented in Proposition 5), feasible solutions were obtained for all the testing networks, and the experimental results obtained have an average gap with respect to the lower bound of 7.01%.

It can be proved that (it is out of the scope of this chapter) increasing k, the following inequality is fulfilled:

$$\frac{c^{k-1}_{opt}}{c^k_{opt}} \leq 1 + \text{floor}\left(\frac{n_C}{k}\right) \cdot \left(\frac{1}{k + n_T}\right) \cdot \left(\frac{c_{max}}{c_{min}} - 1\right) \tag{16}$$

Table 2 shows the results obtained. Despite the bound was not good in these cases (due the heterogeneity of costs of the lines), it can help us in some cases to answer the following question: how much can be saved with a higher k?

Name	n_T	n_C	c_{min}	c_{max}	$\frac{c_{opt^1}}{c_{opt^2}}$	$1 + floor\left(\frac{n_C}{2}\right)\left(\frac{1}{2+n_T}\right)\left(\frac{c_{max}}{c_{min}} - 1\right)$
i080-112	7	72	85	209	1.166401	5.997385619
i080-115	7	72	86	302	1.1004114	10.325581395
i160-015	6	153	86	300	1.223712	23.639534884

Table 2.
Relation between optimal solutions of ANDP$^{(\le 1)}$ *and* ANDP$^{(\le 2)}$.

Set	Name	|V|	|E|	|T|	c_{opt^2}	$t_{c_{opt^2}}$	$LB_{SSR^{(2)}}$	$t_{LB_{SSR^{(2)}}}$	$LBGAP_{SSR^{(2)}}$
I080	i080-001	80	120	6	2187	0	1698	0	28.8%
I080	i080-011	80	350	6	1499	6.04	1307	0.27	14.69%
I080	i080-012	80	350	6	1497	5.33	1486	0.16	0.74%
I080	i080-013	80	350	6	1383	8.20	1000	0.92	38.3%
I080	i080-014	80	350	6	1505	4.89	1211	0.25	24.28%
I080	i080-111	80	350	8	2159	3.09	1982	0.45	8.93%
I080	i080-112	80	350	8	1887	1812	1501	7.52	25.72%
I080	i080-113	80	350	8	1884	1809	1591	393.8	18.42%
I080	i080-114	80	350	8	2099	44.81	1988	6.65	5.58%
I080	i080-115	80	350	8	1969	479.8	1496	15.41	31.62%
I080	i080-233	80	160	16	4564	361.1	3997	6.75	14.19%
I160	i160-011	160	812	7	1875	45.67	1399	2.17	34.02%
I160	i160-012	160	812	7	1891	8.83	1502	1.13	25.9%
I160	i160-013	160	812	7	1862	6.58	1381	1.81	34..83%
I160	i160-014	160	812	7	1991	6.06	1783	0.86	11.67%
I160	i160-015	160	812	7	1864	70.28	1793	6.21	3.96%
PUC	cc3-4p	64	288	8	2553	79.37	2177	2.54	17.27%
PUC	cc3-4u	64	288	8	25	80.04	21	5.18	19.05%
Average								19.89%	

Table 3.
Lower bounds obtained to ANDP$^{(\le 2)}$ *by applying Dynamic Programming with State-Space Relaxation.*

Table 3 shows the results obtained by applying the recursion presented in Proposition 6. As before the first column contains the names of the original SteinLib classes with the name of the customized instance. The entries from left to right are:

- The size of the selected instance in terms of number of nodes, edges, and terminal sites, respectively
- The cost of a global optimal solution of $\Gamma^{(\le 2)}_{ANDP}\left(c^2_{opt}\right)$
- The execution time, in seconds, for $c^2_{opt}\left(t_{c^2_{opt}}\right)$
- A lower bound for the cost of a global optimal solution of $\Gamma^{(\le 2)}_{ANDP}$ by applying Dynamic Programming with State-Space Relaxation (presented in Proposition 6) $(LB^{(2)}_{SSR})$

- The execution time, in seconds, for $LB^{(2)}_{SSR}\left(t_{LB^{(2)}_{SSR}}\right)$

- The gap of the cost for a global optimal solution of $\Gamma^{(\leq 2)}_{ANDP}\left(c^2_{opt}\right)$ with respect to the lower bound $LB^{(2)}_{SSR}$; $LB\_GAP^{(2)}_{SSR}$

The $LB\_GAP^{(2)}_{SSR}$ is computed as

$$LB\_GAP^{(2)}_{SSR} = 100 \times \frac{c^2_{opt} - LB^{(2)}{}_{SSR}}{LB^{(2)}{}_{SSR}} \tag{17}$$

In general, the gaps related to the lower bounds were low. The $r_i$to each terminal site and concentrator site were distinct integers chosen from $\{1, \ldots |S_T \cup S_C|\}$. This lower bound can be increased by modifying the state-space through the application of subgradient optimization to r_i. As future work, it is possible to incorporate the method for a better choice of r_i.

It can be noticed that the execution times of computing global optimal solution costs were much longer than using Dynamic Programming with State-Space Relaxation.

6. Conclusions

The implementation of the algorithms was tested on a number of different problems with heterogeneous characteristics. In particular, a set of ANDP instances transforming 18 SPG instances extracted from SteinLib was built. The optimal values for the selected SPG instances are lower bound for the corresponding ANDP. The solutions found by the algorithm were, in average, 21% and 7% lower than the mentioned bounds in $\text{ANDP}^{(\leq 1)}$ and $\text{ANDP}^{(\leq 2)}$, respectively. It is reasonable supposing that the gaps related to the global optimum of the ANDP instances be even lower since the feasible solutions of the ANDP that are also feasible solutions of the original SPG, but not reciprocally. In this sense, remember that in any ANDP instance generated, all the edges between pairs of terminal nodes were deleted (because in our ANDP such connections are not allowed) having the additional constraint that the terminal nodes must have degree one in the solution.

Besides, a Dynamic Programming with State-Space Relaxation algorithm was developed which can give a lower bound in polynomial time. The average gaps with respect to the global optimal solution costs were lower than 20%.

Notice that, as expected, the execution times of the proposed algorithms are strongly dependent on the number of sites, edges, and terminal sites. To sum up, as far as the authors are concerned, the results obtained with the recurrences above are very good, considering that computing the global optimal solution of an $\text{ANDP}^{(\leq 2)}$ is a NP-hard problem.

Author details

Franco Robledo[1], Pablo Romero[1], Pablo Sartor[2*], Luis Stábile[1] and Omar Viera[1]

1 Instituto de Computación, Facultad de Ingeniería, Universidad de la República, Montevideo, Uruguay

2 Departamento de Análisis de Decisiones, IEEM Business School, Universidad de Montevideo, Montevideo, Uruguay

*Address all correspondence to: psartor@um.edu.uy

References

[1] Xie S, Ouyang Y. Reliable service systems design under the risk of network access failures. Transportation Research Part E: Logistics and Transportation Review. 2019;**122**(1): 1-13. DOI: 10.1016/j.tre.2018.11.002

[2] Lee Y, Kim Y, Park G. An access network design problem with end-to-end QoS constraints. Omega. 2014; **48**(1):36-48

[3] Zhang J, Sun X, Wandelt S. HUBBI: Iterative network design for incomplete hub location problems. Computers and Operations Research. 2019;**104**(1): 394-414. DOI: 10.1016/j.cor.2018.09.011

[4] Ljubic I, Putz P, Salazar-González J. A MIP-based approach to solve the prize-collecting local access network design problem. European Journal of Operational Research. 2014;**253**(3): 727-739

[5] Priem M, Priem F. Ingénierie des WAN (text in French). Paris: Dunod InterEditions; 1999

[6] Stoer M. Design of Survivable Networks. Vol. 1532. Berlin Heidelberg: Springer-Verlag; 1992

[7] Robledo F. GRASP heuristics for a Wide Area Network design [PhD thesis]. Universidad de la República Oriental del Uruguay and Université de Rennes I; 2005

[8] Christofides N, Mingozzi A, Toth P. State-space relaxation procedures for the computation of bounds to routing problems. Networks. 1981;**11**(2): 145-164. DOI: 10.1002/net.3230110207

[9] Mingozzi A. State space relaxation and search strategies in dynamic programming. In: Proceedings of the 5th International Symposium on Abstraction, Reformulation and Approximation. London, UK: Springer-Verlag; 2002. p. 51. Available from: http://portal.acm.org/citation.cfm?id=645848.758271

[10] Koch T, Martin A, Voß S. SteinLib: An updated library on Steiner tree problems in graphs. Technical Report ZIB-Report 00-37. Berlin: Konrad-Zuse-Zentrum für Informationstechnik Berlin; 2000. Available from: http://elib.zib.de/steinlib

Chapter 5

The Graphs for Elliptic Curve Cryptography

Ruma Kareem K. Ajeena

Abstract

The scalar multiplication on elliptic curves defined over finite fields is a core operation in elliptic curve cryptography (ECC). Several different methods are used for computing this operation. One of them, the binary method, is applied depending on the binary representation of the scalar v in a scalar multiplication vP, where P is a point that lies on elliptic curve E defined over a prime field F_p. On the binary method, two methodologies are performed based on the implementation of the binary string bits from the right to the left (RLB) [or from the left to the right (LRB)]. Another method is a nonadjacent form (NAF) which depended on the signed digit representation of a positive integer v. In this chapter, the graphs and subgraphs are employed for the serial computations of elliptic scalar multiplications defined over prime fields. This work proposed using the subgraphs H of the graphs G or the (simple, undirected, directed, connected, bipartite, and other) graphs to represent a scalar v directly. This usage speeds up the computations on the elliptic scalar multiplication algorithms. The computational complexities of the proposed algorithms and previous ones are determined. The comparison results of the computational complexities on all these algorithms are discussed. The experimental results show that the proposed algorithms which are used the sub-graphs H and graphs G need to the less costs for computing vP in compare to previous algorithms which are employed the binary representations or NAF expansion. Thus, the proposed algorithms that use the subgraphs or the graphs to represent the scalars v are more efficient than the original ones.

Keywords: ECC, scalar multiplication, BRL, BLR, NAF, graphs, subgraphs, computational complexity

1. Introduction

The scalar multiplication on elliptic curves defined over finite fields is considered as a central and most time-consuming operation in elliptic curve cryptography (ECC) [1–7]. Different methods are used for computing the scalar multiplication such as the binary method, nonadjacent form, and others [8–15]. The binary method is applied depending on the binary representation of the scalar v in a scalar multiplication vP, where P is a point that lies on elliptic curve E defined over a prime field F_p. On the binary method, two methodologies are performed based on the implementation of the binary string bits from the right to the left (RLB) [or from the left to the right (LRB)], whereas the nonadjacent form (NAF) depends on the signed digit representation of a positive integer v [1].

IntechOpen

In this chapter, the computation of the scalar multiplication vP on elliptic curve E defined over a prime field F_p has been done using the (undirected or directed) graph and (undirected or directed) subgraph. These graph and subgraph are used to represent the scalar v in two ways. The first one is the binary representation and the second one is the sign digit representation.

Also, the l-tuple of the elliptic scalar multiplications is computed using the proposed generalized binary methods (GRLB) and (GLRB) and GNAF. The computational complexities of the proposed algorithms and previous ones are determined. The comparison results of the computational complexities on all these algorithms are discussed. Several experimental results showed that the proposed algorithms which are used the graphs G need to the less costs for computing vP in compare to previous algorithms which are employed the binary representations or NAF expansion. Therefore, the proposed algorithms that use the subgraphs or the graphs to represent the scalars v are more efficient than the original ones.

This chapter is organized as follows: Section 2 presents the vector representation of the graph. Section 3 discusses the matrix representation of the graph. Section 4 includes the binary methods of the elliptic scalar multiplication which are the right-to-left binary and left-to-right binary representations. Section 5 explains the non-adjacent form method, whereas Section 6 discusses the graphic binary methods of the elliptic scalar multiplications. Section 7 displays the digraphic NAF method. Section 8 presents the subgraphs for computing the elliptic scalar multiplication. Section 9 determines the computational complexities on the original elliptic scalar multiplication methods. Section 10 shows the computational complexity for serial computing l-tuple of the scalar multiplications. The computational complexity of the graphic elliptic scalar multiplication methods is explained in Section 11. Section 12 illustrates the computational complexity comparison on the serial and graphic computation methods. Finally, Section 13 draws the conclusions.

2. The vector representation of the graph

Suppose G is a graph as shown in **Figure 1**.

A graph G has four vertices and five edges e_1, e_2, e_3, e_4, and e_5. A subgraph H (and any other subgraphs) of G is represented by a 5-tuple.

This means that $E = (e_1, e_2, e_3, e_4, e_5)$ such that

$$e_i = 1, \text{ if } e_i \text{ is in } H,$$

$$e_i = 0, \text{ if } e_i \text{ is not in } H.$$

The subgraphs H_1 and H_2 in **Figure 1** can be represented by (1,0,1,0,1) and (0,1,1,1,0), respectively. Here, there are $2^5 = 32$ possible cases for 5-tuples which

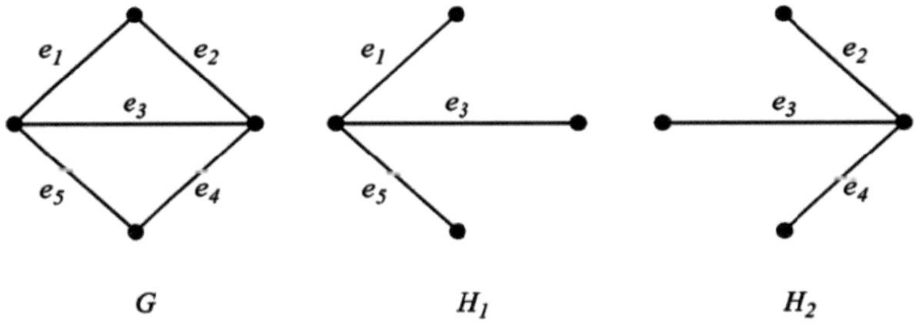

Figure 1.
The subgraphs H_1 and H_2 of the graph G [16].

correspond to 32 subgraphs. Among them are the (0,0,0,0,0) and (1,1,1,1,1) which represent a null graph and a graph G itself, respectively [16].

3. The matrix representation of the graph

Suppose G is any undirected graph that is formed by two finite sets V and E, which are called the vertices and edges, respectively. In other words, $V = \{v_1, v_2, \dots, v_l\}$ and $E = \{e_1, e_2, \dots, e_m\}$. The matrix representation $A(G) = (e_{ij})_{l\times m}$ on graph G has been defined by

$$A(G) = \begin{array}{c|c} & \\ v_1 & \\ v_2 & \\ \vdots & \\ v_l & \end{array} \begin{bmatrix} e_{1_1} & e_{2_1} & e_{3_1} & \dots & e_{m_1} \\ e_{1_2} & e_{2_2} & e_{3_2} & \dots & e_{m_2} \\ \vdots & \vdots & \vdots & \vdots & \vdots \\ e_{1_l} & e_{2_l} & e_{2_l} & \dots & e_{m_l} \end{bmatrix} \tag{1}$$

with l rows corresponding to the l vertices v_i and the m columns corresponding to the m edges e_i. Whereas the incidence matrix of a connected digraph can be defined by $A = (e_{ij})_{l\times m}$, where $e_{ij} \in \{0, \mp 1\}$. In other words, if j^{th} edge is incident out of ith vertex, then $e_{ij} = 1$, while $e_{ij} = -1$, if j^{th} edge is incident into i^{th} vertex and if j^{th} edge is neither incident out nor incident into i^{th} vertex, then $e_{ij} = 0$ [16, 17].

4. The binary methods for the elliptic scalar multiplication

Two methods for computing the scalar multiplication vP have been created based on using the binary representation of a scalar v. One of them is called the right-to-left binary (RLB) method, and another one is called left-to-right binary (LRB) method [1, 9, 10]. These methods depend on the basic repeated-square-and multiply methods for exponentiation with additive version. Using the RLB method, the process of v-bits starts from the right to the left, whereas the v-bits processing starts from the left to the right using the LRB method. The RLB and LRB methods are discussed mathematically as follows.

4.1 The right-to-left binary method

Suppose E is an elliptic curve defined over a prime field F_p. The equation of E is given by $E: y^2 = x^3 + ax + b \pmod p$. Let $P = (x, y)$ be a generator point that lies on E which has a (large) prime order n. Choosing v to compute vP can be done from the range $[1, n-1]$. So, it should first write v in a binary representation string $(e_{t-1}, \dots, e_1, e_0)_2$. The starting will be happened with a point Q in E (F_p), (that is, $Q = \infty$). With the i index that takes the values $0, 1, \dots, t-1$, the computation of $Q = Q + P$ can be done if $e_i = 1$. After then, the value $2P$ is computed and plugging $2P$ by P. The processing continues until the last value $t - 1$. Therefore, the last computed value of a point Q is the scalar multiplication point vP [1]. The summary of the RLB method can be given in the following algorithm.

Algorithm 4.1 The RLB algorithm

Input: A scalar v in [1, n-1] and a point P in $E(F_p)$.
Output: A scalar multiplication vP.

1. Write down a scalar v as a binary string $v = (e_{t-1}, ..., e_1, e_0)_2$.

2. $Q = \infty$.

3. For $i = 0,1,..., t-1$ do

   3.1 If $e_i = 1$ then $Q = Q + P$.

   3.2 Compute $P = 2P$.

   3.3 Else compute $P = 2P$.

   3.4 End if

4. End for

5. Return $Q = vP$.

4.2 The left-to-right binary method

With the same parameters E, P, n, and v which are used in the RLB method, the computation of vP using the LRB method can be done easily. A scalar v can be written in a binary representation string $(e_{t-1}, ..., e_1, e_0)_2$. Let us start with a point Q in $E(F_p)$, where $Q = \infty$. With the i index which takes the values $t-1,..., 1, 0$, then the computation of $2Q$ can be done and plugged into Q. After then, the value $Q = Q + P$ is computed. The processing continues until the last value 0. Therefore, the last computed value of a point Q is the scalar multiplication point vP. The LRB method can be summarized in Algorithm (4.2) [1].

Algorithm 4.2 The LRB algorithm

Input: A scalar v in [1, n-1] and a point P in $E(F_p)$.
Output: A scalar multiplication vP.

1. Write down a scalar v as a binary string $v = (e_{t-1},..., e_1, e_0)_2$.

2. $Q = \infty$.

3. For $i = t-1,...,1, 0$ do

   3.1 Compute $Q = 2Q$.

   3.2 If $e_i = 1$ then $Q = Q + P$.

   3.3 Else go to step (3.4).

   3.4 End if

4. End for

5. Return $Q = vP$.

5. The non-adjacent form for the elliptic scalar multiplication

The motivation to use the signed digit representation of a scalar v, in a scalar multiplication vP, is the computation of the subtraction and addition of the points lying on elliptic curve E which has the same efficient. A signed digit representation of v is given by $v = \sum_{i=0}^{l-1} e_i 2^i$, where $e_i \in \{0, \pm 1\}$ will be explained in this section with more details. The signed digit representation forms the nonadjacent form (NAF) [1, 9, 10] which is given in the next algorithm.

Algorithm 5.1 The NAF computation of a positive integer

Input: A positive integer v in $[1, n\text{-}1]$.
Output: The expansion NAF (v).

1. $i \leftarrow 0$.

2. While $v \geq 1$ do

   2.1 If v is odd then $e_i \leftarrow 2 - (v \bmod 4)$,

   $v \leftarrow v - e_i$;

   2.2 Else: $e_i \leftarrow 0$.

   2.3 End if

3. $v \leftarrow v / 2$, $i \leftarrow i + 1$.

4. End while

5. Return $(e_{i-1}, \ldots, e_1, e_0)$.

The computation of a scalar multiplication vP by employing the NAF algorithm can be done using the following algorithm:

Algorithm 5.2 The NAF method for computing the scalar multiplication

Input: A positive integer v in $[1, n\text{-}1]$ and $P \in E(F_p)$.
Output: A scalar multiplication vP.

1. Algorithm (5.1) uses to compute NAF(v).

2. $Q \leftarrow \infty$.

3. For $i = \mathrm{t} - 1, \ldots, 1, 0$ do

   3.1 $Q \leftarrow 2Q$.

   3.2 If $e_i = 1$ then $Q \leftarrow Q + P$.

   3.3 ElseIf $e_i = -1$ then $Q \leftarrow Q - P$.

3.4 Else go to step (3.5).

3.5 End if

4. End for

5. Return ($Q = vP$).

6. The graphic methods for the elliptic scalar multiplications

This section discusses the generalization on the binary methods and NAF to compute l-tuple of the scalar multiplications on elliptic curve E defined over prime field Fp. This generalization employed the simple undirected and directed graphs.

6.1 The graphic right-to-left binary (GRLB) method

Suppose Ec is an elliptic curve defined over a prime field F_p [1–7]. The equation of Ec is given by

$$Ec : y^2 = x^3 + ax + b \pmod{p}. \tag{2}$$

Let $P = (x, y)$ be a point that lies on Ec which has a (large) prime order r. Let $G(V, E)$ be a simple (or multigraph or others) graph, where V is a vertex set and E is an edge set. The matrix representation $A(G)$ on $G(V, E)$ is defined as given in Eq. (1). Directly from the rows of the matrix $A(G)$, the binary representation strings $\left(e_{(m-1)_l}, \dots, e_{1_l}, e_{0_l}\right)_2$ are obtained. The starting will happen with an elliptic point Q_1 which belongs to $E(F_p)$, where $Q_1 = \infty$. With the i index which takes the values $0_1, 1_1, \dots, (m-1)_1$ in the first row of $A(G)$, the computation of $Q_1 = Q_1 + P$ can be done if $e_{i_1} = 1$. After then, the value $2P$ is computed and plugging it by P. The processing on the first row continues until the last value $m - 1$. Therefore, the last computed value of a point Q_1 is the value of the first scalar multiplication point v_1P in l-tuple $\langle vP \rangle$. In similar way, the processing on others rows can be done. The summary of the GRLB method can be given in the following algorithm:

Algorithm 6.1 The GRLB method

Input: A graph $G(V, E)$, $P \in E(F_p)$, l and m, where l and m are the order and size of a graph G, respectively.
Output: The m-tuple of the scalar multiplications $\langle vP \rangle = \langle v_1P, \dots, v_lP \rangle$.

1. Write down the matrix representation $A(G)$ of the graph $G(V, E)$.

2. Directly determine the binary representation strings $v_j = \left(e_{(m-1)_j}, \dots, e_{1_j}, e_{0j}\right)_2$ from $A(G)$

3. For $j = 1, 2, \dots, l$.

4. $\quad Q_j \leftarrow \infty$.

5. $\quad$ For $i = 0_j : (m-1)_j$ do

5.1 If $e_{i_j} = 1$ then $Q_j = Q_j + P$.

5.2 Else go to step (6).

5.3 End if

6. Compute $P \leftarrow 2P$.

7. End for

8. Return $(Q_j = v_j P)$.

9. End for

10. Return $(\langle Q \rangle = \langle vP \rangle = \langle v_1 P, v_2 P, \ldots, v_l P \rangle)$.

6.2 The implementation results on the GRLB method

With different kinds of graphs which are given in **Figure 2**, the matrix representations of the graphs have been computed by $A(G_a)$, $A(G_b)$, $A(G_c)$, and $A(G_d)$, respectively.

$$A(G_a) = \left|\begin{array}{c} v_1 \\ v_2 \\ v_3 \\ v_4 \\ v_5 \end{array}\right. . \begin{bmatrix} 1 & 0 & 1 & 1 & 0 & 0 & 1 \\ 1 & 1 & 0 & 0 & 0 & 0 & 0 \\ 0 & 0 & 1 & 0 & 1 & 0 & 0 \\ 0 & 0 & 0 & 1 & 0 & 1 & 0 \\ 0 & 0 & 0 & 1 & 1 & 1 & 1 \end{bmatrix}, \quad A(G_b) = \left|\begin{array}{c} v_1 \\ v_2 \\ v_3 \\ v_4 \end{array}\right. . \begin{bmatrix} 1 & 0 & 1 & 0 & 0 & 1 \\ 1 & 1 & 0 & 1 & 0 & 0 \\ 0 & 0 & 0 & 1 & 1 & 1 \\ 0 & 1 & 1 & 0 & 1 & 0 \end{bmatrix}$$

$$A(G_c) = \left|\begin{array}{c} v_1 \\ v_2 \\ v_3 \\ v_4 \\ v_5 \\ v_6 \end{array}\right. . \begin{bmatrix} 1 & 0 & 1 & 1 & 0 & 0 & 0 & 0 \\ 1 & 1 & 0 & 0 & 0 & 0 & 0 & 0 \\ 0 & 0 & 1 & 1 & 1 & 0 & 0 & 0 \\ 0 & 1 & 0 & 0 & 1 & 1 & 1 & 0 \\ 0 & 0 & 0 & 0 & 0 & 1 & 0 & 1 \\ 0 & 0 & 0 & 0 & 0 & 0 & 1 & 1 \end{bmatrix} \text{ and } A(G_d) = \left|\begin{array}{c} v_1 \\ v_2 \\ v_3 \\ v_4 \\ v_5 \\ v_6 \end{array}\right. . \begin{bmatrix} 1 & 1 & 0 & 0 & 0 & 0 & 0 & 1 & 0 \\ 0 & 0 & 1 & 1 & 0 & 0 & 1 & 0 & 0 \\ 0 & 0 & 0 & 0 & 1 & 1 & 0 & 0 & 1 \\ 0 & 0 & 1 & 0 & 1 & 0 & 0 & 1 & 0 \\ 0 & 1 & 0 & 0 & 0 & 0 & 1 & 0 & 1 \\ 1 & 0 & 0 & 1 & 0 & 1 & 0 & 0 & 0 \end{bmatrix}$$

The l-tuple computations of the scalar multiplications that correspond to these graphs are shown in **Table 1**.

6.3 The graphic left-to-right binary method

With the same parameters p, E, P, G, and V which are used in the GRLB method, the computations of l-tuple $\langle vP \rangle$ using the GLRB method can be done easily. The scalars $v_1, \ldots, v_n$ can be written in the binary representation strings $\left(e_{(m-1)_j}, \ldots, e_{1j}, e_{0_j}\right)_2$, for j = 1, 2, …, l, directly from the matrix representation $A(G)$ of G. Let us start with a point Q_1 in $E(F_p)$, where $Q_1 = \infty$. With the i index which takes the values $(m-1)_1, \ldots, 1_1, 0_1$, then the computation of $2Q_1$ can be done and plugged into Q_1. After then, the value $Q_1 = Q_1 + P$ is computed. The processing continues until the last value 0_1. Therefore, the last computed value of a point Q_1 is the first scalar multiplication point in an l-tuple $\langle vP \rangle$. Similarly, the processing on others rows can be computed. The GLRB method can be summarized in Algorithm (6.2).

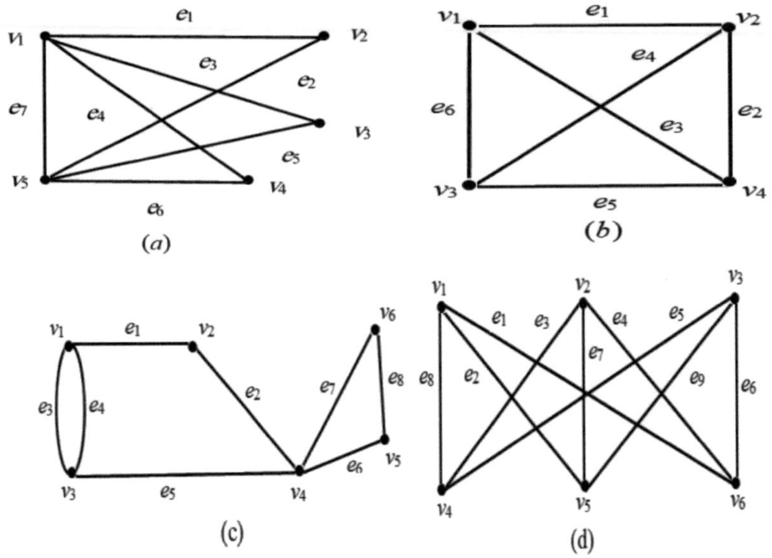

Figure 2.
Different kinds of graphs [16].

P	E (a,b)	N	Generator point	G	G (l,m)	⟨vP⟩
101	E (10,2)	109	P = (68,14)	G_a	G_a (5,7)	$\langle v_1P, v_2P, v_3P, v_4P, v_5P\rangle = \langle(14,19),(91,66),(44,68),(5,51),(93,4)\rangle$
61	E (4,1)	67	P = (24,14)	G_b	G_b (4, 6)	$\langle v_1P, v_2P, v_3P, v_4P\rangle = \langle(0,60),(4,52),(43,21),(0,1)\rangle$
191	E (7,2)	193	P = (41,91)	G_c	G_c (6, 8)	$\langle v_1P, v_2P, v_3P, v_4P, v_5P, v_6P\rangle = \langle(24,137),(41,100),(43,113),(18,109),(16,114),(105,86)\rangle$
449	E (2,2)	467	P = (50,27)	G_d	G_d (6, 9)	$\langle v_1P, v_2P, v_3P, v_4P, v_5P, v_6P\rangle = \langle(93,281),(405,104),(96,20),(266,382),(236,399),(31,391)\rangle$

Table 1.
The experimental results of the l*-tuple of the scalar multiplications that correspond to the graphs* G_a, G_b, G_c, *and* G_d.

Algorithm 6.2 The GLRB method

Input: A graph $G(V,E)$, $P \in E(F_p)$, l and m.
Output: The l-tuple of the scalar multiplications $\langle vP\rangle = \langle v_1P, \ldots, v_lP\rangle$.

1. Write down the matrix representation $A(G)$ of the graph $G(V,E)$.
2. Directly determine the binary representation strings $v_j = \left(e_{(m-1)_j}, \ldots, e_{1_j}, e_{0_j}\right)_2$, for j=1,2,..., l from $A(G)$.
3. For $j = 1, 2, \ldots, l$.
4. $\quad Q_j \leftarrow \infty$.
5. $\quad$ For $i = (m-1)_j : 0_j$ do

5.1 Compute $Q_j = 2Q_j$.

5.2 If $e_{i_j} = 1$ then $Q_j = Q_j + P$.

5.3 Else go to Step (5.4).

5.4 End if

6. End for

7. Return ($Q_j = v_j P$).

8. End for

9. Return ($\langle Q \rangle = \langle vP \rangle = \langle v_1 P, v_2 P, \ldots, v_l P \rangle$).

7. The digraphic NAF for the elliptic scalar multiplication

The signed digit representation of an l -tuple $\langle\, v \rangle$ of scalars v_j, which are used to compute an l-tuple $\langle\, vP \rangle$ of the scalar multiplications $v_j P$, can be represented directly from the digraphs. The signed digit representations of v_j are given by $v_j = \sum_{i=0}^{l-1} e_{i_j} 2^{i_j}$, where $e_{i_j} \in \{0, \pm 1\}$. The signed digit representations form the generalized nonadjacent form (GNAF). These representations are computed using the following algorithm:

Algorithm 7.1 The GNAF computation of an l-tuple of the positive integers

Input: An l-tuple of positive integers v_j.
Output:$\langle\, NAF_S(v) \rangle = \langle\, NAF_S(v_1), NAF_S(v_2), \ldots, NAF_S(v_l) \rangle$.

1. Determine v_j, j = 1, 2, ...,l and $\left(e_{1j}, e_{2j}, \ldots, e_{m_j}\right)$in any digraph G.

2. For j = 1, 2, ..., l.

3. For i = 1,...,m.

4. If v_s is an incident out of v_t, where $s, t \in j$

5. then $e_{i_j} = 1$.

6. Elseif v_s is an incident into v_t

7. then $e_{ij} = -1$.

8. Else there is no edge between v_s and v_t.

9. then $e_{i_j} = 0$.

10. End if

11. End For

12. Return $\left(e_{1j}, e_{2j}, \ldots, e_{m_j}\right)$.

13. End For

14. Return $NAF(v_j) = \left(e_{m_j}, \ldots, e_{2j}, e_{1j}\right)$.

In **Figure 3**, the digraph G has the vertices v_j for $j = 1, 2, 3, 4$ and edges e_m for $m = 1, 2, \ldots, 7$.

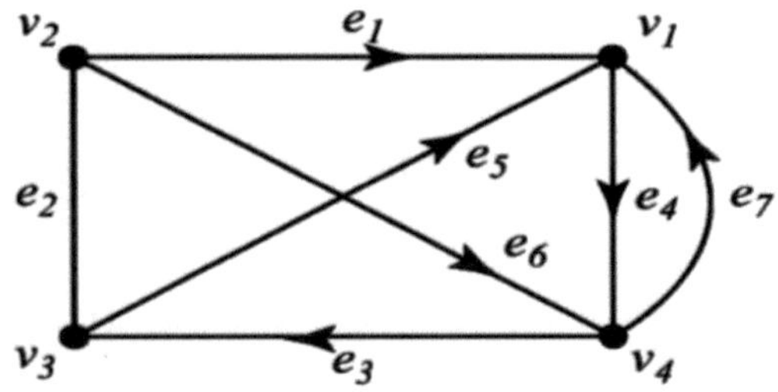

Figure 3.
The digraph has the vertices v_j *for* $j = 1, 2, 3, 4$ *and edges* e_m *for* $m = 1, 2, \ldots, 7$.

The incidence matrix of G that is given in **Figure 3** is

$$A = \begin{array}{c|c} v_1 & \\ v_2 & \\ v_3 & \\ v_4 & \end{array}\begin{bmatrix} -1 & 0 & 0 & 1 & -1 & 0 & -1 \\ 1 & 1 & 0 & 0 & 0 & 1 & 0 \\ 0 & -1 & -1 & 0 & 1 & 0 & 0 \\ 0 & 0 & 1 & -1 & 0 & -1 & 1 \end{bmatrix}.$$

So, the NAF representations of 4-tuple $\langle v_1, v_2, v_3, v_4 \rangle$ are

$$\langle(-1,0,0,1,-1,0,-1),(1,1,0,0,0,1,0),(0,-1,-1,0,1,0,0),(0,0,1,-1,0,-1,1)\rangle.$$

The GNAF method for l-tuple of the scalar multiplications can be performed using Algorithm (7.2).

Algorithm 7.2 The GNAF method for computing l-tuple of the scalar multiplication

Input: The l-tuple of positive integers v_j and $P \in E(F_p)$.
Output: The l-tuple of the scalar multiplications $\langle vP \rangle$.

1. Algorithm (7.1) uses to compute GNAF(v).

2. $Q_j \leftarrow \infty$.

3. For $j = 1, 2, \ldots, l$

4. For $i = t - 1, \ldots, 1, 0$

4.1 $Q_j \leftarrow 2Q_j$.

4.2 If $e_{i_j} = 1$ then $Q_j \leftarrow Q_j + P$.

4.3 Elseif $e_{i_j} = -1$ then $Q_j \leftarrow Q_j - P$.

4.4 Else go to step (4.5).

4.5 End if

5. End for

6. End for

7. Return $\left\langle Q_j = v_j P \right\rangle$.

Using Algorithm (7.2), the final result of 4-tuple of the scalar multiplications is given by

$$\langle v_1 P, v_2 P, v_3 P, v_4 P \rangle = \langle (28, 32), (46, 63), (25, 90), (82, 15) \rangle.$$

8. The subgraphs for the elliptic scalar multiplication

8.1 The binary representations

Suppose G is a graph and H_i, for $i = 1, 2, 3$ are subgraphs as shown in **Figure 4**.
Next algorithm can be applied for determining the binary representation of any subgraph from a given graph.

Algorithm 8.1 The graphic binary representation of a subgraph from a given graph

Input: A graph $G(V, E)$, where $V = (v_1, v_2, \ldots, v_l)$ and $E = (e_1, e_2, \ldots, e_m)$.
Output: The $BR_{subgraph}(v)$.

1. Determine $(v_1, v_2, \ldots, v_k)$ and $(e_1, e_2, \ldots, e_m)$ in any subgraph H of G.

2. $i \leftarrow 0$.

3. For $j = 0{:}\ k$, where $k \leq l$.

4. If there is an edge between v_s and v_t, where $s, t \in j$

5. then $e_i = 1$.

6. Else there is no edge between v_s and v_t.

7. then $e_i = 0$.

8. End if

9. $i \leftarrow i + 1$.

10. Return $BR_{subgraph} = (e_{m-1}, \ldots, e_1, e_0)_2$.

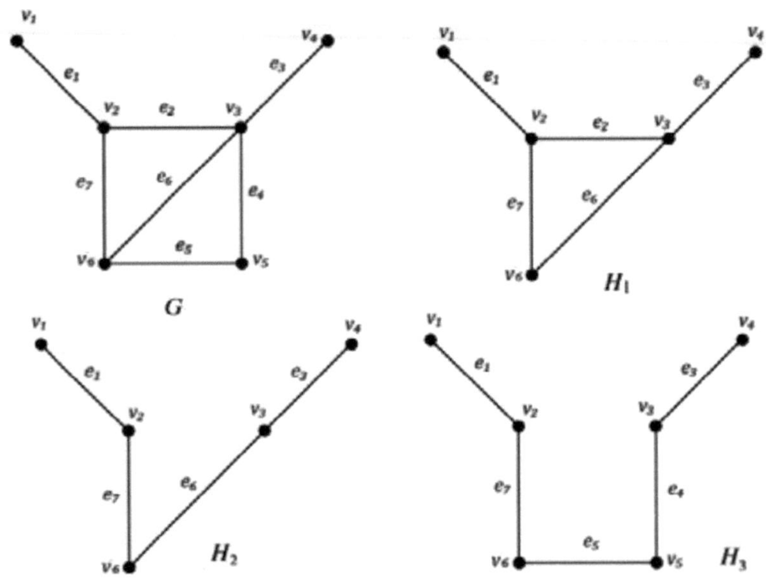

Figure 4.
The subgraphs Hi, for i = 1, 2, 3, for a graph G.

Subgraphs	$(v_1, v_2, \ldots, v_k)$	$(e_1, e_2, \ldots, e_m)$	$BR_{subgraph} = (e_{m-1}, \ldots, e_1, e_0)_2$
H_1	$(v_1, v_2, v_3, v_4, v_6)$	$(e_1, e_2, e_3, e_6, e_7)$	(1, 1, 0, 0, 1, 1, 1)
H_2	$(v_1, v_2, v_3, v_4, v_6)$	(e_1, e_3, e_6, e_7)	(1, 1, 0 ,0, 1, 0, 1)
H_3	$(v_1, v_2, v_3, v_4, v_5, v_6)$	$(e_1, e_3, e_4, e_5, e_7)$	(1, 0, 1, 1, 1, 0, 1)

Table 2.
The experimental results of the binary representations of scalars using subgraphs.

p	E (*a*,*b*)	n	Gen Pt P	Subgraph	$BR_{subgraph} = (e_{m-1}, \ldots, e_1, e_0)_2$	H_iP
191	E (7,2)	193	$P = (41,91)$	H_1	(1, 1, 0, 0, 1, 1, 1)	(80,142)
				H_2	(1, 1, 0 ,0 ,1 , 0, 1)	(0,57)
				H_3	(1, 0, 1, 1, 1, 0, 1)	(36,146)

Table 3.
The experimental results for computing of the scalar multiplications based on using the binary representation of the subgraphs.

The small numerical results based on **Figure 4** can be shown in **Table 2**.

On the binary representations which are found directly from the subgraphs, the scalar multiplications H_iP on elliptic curve E defined over a prime field Fp can be computed using Algorithm (4.1) or (4.2). Some experimental results for computing the scalar multiplications based on using the subgraphs to represent the scalars are given in **Table 3**.

9. The signed digit representations

Suppose G is a digraph and H_i, for $i = 1, 2, 3$, are directed subgraphs as shown in **Figure 5**. Algorithm (8.2) can be used to find the signed digit representation of any subgraph from a given graph.

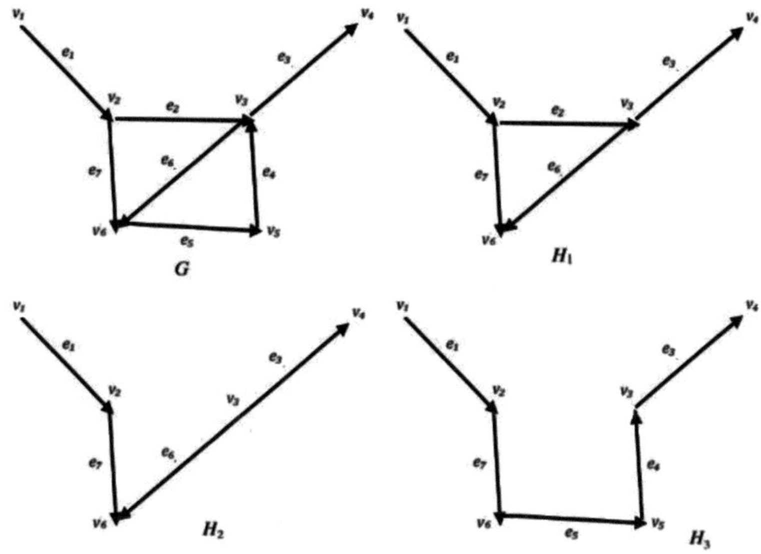

Figure 5.
The directed subgraphs Hi, *for i = 1, 2, 3, 4, for a digraph* G.

Algorithm 8.2 The di-subgraph signed digit representation of the positive integers

Input: A directed graph $G(V, E)$, where $V = (v_1, v_2, \ldots, v_l)$ and $E = (e_1, e_2, \ldots, e_m)$.
Output: The $\text{SDR}_{\text{subgraph}}(v)$.

1. Determine $(v_1, v_2, \ldots, v_k)$ and $(e_0, e_1, \ldots, e_{m-1})$ in any subgraph H of G.

2. $i \leftarrow 0$.

3. For j = 0: k, where k l.

4. If v_s is an incident out of v_t, where $s, t \in j$

5. then $e_i = 1$.

6. Elseif v_t is an incident into v_s

7. then $e_i = -1$.

8. Else there is no edge between v_s and v_t.

9. then $e_i = 0$.

10. End if

11. End for

11. $i \leftarrow i+1$.

12. Return $(e_{m-1}, \ldots, e_1, e_0)$.

Subgraphs	l-tuple $\langle v \rangle$	l-tuple $\langle NAF_S(v) \rangle$
H_1	$\langle v_1, v_2, v_3, v_6 \rangle$	$\langle (1,0,0,0,0,0,0), (-1,1,0,0,0,0,1), (0,-1,1,0,0,1,0), (0,0,0,0,0,-1,-1) \rangle$
H_2	$\langle v_1, v_2, v_3, v_4, v_6 \rangle$	$\langle (1,0,0,0,0,0,0), (-1,0,0,0,0,0,1), (0,0,1,0,0,1,0), (0,0,-1,0,0,0,0), (0,0,0,0,0,-1,-1) \rangle$
H_3	$\langle v_1, v_2, v_3, v_4, v_5, v_6 \rangle$	$\langle (1,0,0,0,0,0,0), (-1,0,0,0,0,0,1), (0,0,1,-1,0,0,0), (0,0,-1,0,0,0,0), (0,0,0,1,-1,0,0), (0,0,0,0,1,0,-1) \rangle$

Table 4.
The experimental results for sign digit representing l*-tuple of the scalars using the subgraphs.*

P	P	Directed subgraphs	l-tuple $\langle v \rangle$	$\langle vP \rangle$
191	$P = (41,91)$	H_1	$\langle v_1, v_2, v_3, v_6 \rangle$	$\langle (133,91), (171,71), (132,144), (16,77) \rangle$
		H_2	$\langle v_1, v_2, v_3, v_4, v_6 \rangle$	$\langle (133,91), (17,91), (177,186), (177,5), (16,77) \rangle$
		H_3	$\langle v_1, v_2, v_3, v_4, v_5, v_6 \rangle$	$\langle (133,91), (17,91), (49,23), (177,5) (79,97), (105,86) \rangle$

Table 5.
The experimental results for computing l*-tuple of the scalar multiplications based on using the subgraphs.*

The computational results based on **Figure 5** and using Algorithm (8.2) are given in **Table 4**. With the signed digit representations which are given in **Table 4**, the l-tuple of the scalar multiplications on elliptic curve E defined over a prime field F_p can be computed. Some experimental results for computing the l-tuple of the scalar multiplications based on using the directed subgraphs to represent the scalars are given in **Table 5**.

10. The computational complexity on the elliptic scalar multiplication methods

This chapter discusses the problems of the computational complexities which are determined depending on the account operations. These operations are the elliptic curve operations, namely, the addition A and doubling D on the points which lie on elliptic curve E defined over a prime field F_p. Also, the finite field operations which are field inversion I, field multiplication M and a field squaring S. The computational complexity problems are determined first of the original binary methods and NAF for computing the scalar multiplications on E. The computational complexities of the proposed methods which are dependent on the graphs and subgraphs are determined as well.

10.1 The computational complexity of the binary methods

Let $\#E\ (F_p) = n$, where n is prime number and it is the nearest number to prime p. A point P in E(Fp) which has order n. Suppose v is a scalar such as v is a randomly selected integer from the interval $[1, n-1]$. The binary representation of v is denoted $(e_{m-1} \ldots e_2.e_1.e_0)_2$ where $m \approx t = \log_2 p$.

The computational complexity of Algorithm (4.1) or (4.2) is roughly $t/2$ point additions and t point doublings, which is denoted by

$$\frac{t}{2}A + tD, \tag{3}$$

in addition to the time of binary representation which is approximately t/2d and t/2S, where d and S are normal addition and squaring. Using Lemmas (6.1) and (6.2) in [18, 19], the points addition A and doubling D can be re-expressed by 1I + 2 M + 1S and 1I + 2 M + 2S, respectively. In other words, the computational complexity of Algorithm (4.1) or (4.2) is expressed in terms of field operations by.

$$3tS + 3tM + 1.5tI + 0.5td. \tag{4}$$

Several computational complexity results to compute a scalar multiplication by applying the binary method are given in **Table 6**.

10.2 The computational complexity of the NAF

With same the multiplier v which belongs to the interval [1,n − 1], the computational complexity to compute a scalar multiplication vP using the NAF is given by

$$D + \frac{t}{3}A + tD = \frac{t}{3}A + (t+1)D. \tag{5}$$

In Eq. (5), D in the first term is the cost of NAF to represent a positive integer v, $t/3A$ + tD is the cost of computing a scalar multiplication vP using NAF method, and t is the length of the NAF string. In other words, the running time of Algorithm (5.1) is expressed in terms of field operations by

$$t/3(1I + 2\,M + 1S) + (t+1)(1I + 2\,M + 2S) = ((t/3) + t + 1)I + ((2/3)t + 2\,t + 2)M + ((t/3) + 2\,t + 2)S. \tag{6}$$

P	*E* (*a*,*b*)	*n*	Gen. pt. *P*	*vP*	Bin. representation	Comp. complexity
101	*E* (10,2)	109	(68,14)	93*P*	(1, 0, 1, 1, 1, 0, 1)	21S + 21 M + 10.5I + 3.5d
61	*E* (4,1)	67	(24,14)	23*P*	(1, 0, 1, 1, 1)	15S + 15 M + 7.5I + 2.5d
113	E (12,4)	103	(52,41)	39*P*	(1, 0, 0, 1, 1, 1)	18S + 18 M + 9I + 4.5d
149	E (13,1)	167	(32,133)	13*P*	(1, 1, 0, 1)	12S + 12 M + 6I + 2d
1031	E (15,7)	1061	(217,808)	281*P*	(1, 0, 0, 0, 1, 1, 0, 0, 1)	27S + 27 M + 13.5I + 4.5d

Table 6.
The experimental results of the computational complexity for the scalar multiplications using the binary method.

P	*E* (*a*,*b*)	*N*	Gen. pt. *P*	*vP*	NAF. rep.	Comp. complexity
101	*E* (10,2)	109	(68,14)	93*P*	(1, 0, −1, 0, 0, −1, 0, 1)	11.6I + 23.3 M + 20.6S
61	*E* (4,1)	67	(24,14)	23*P*	(1, 0, −1, 0, 0, −1)	9I + 18 M + 16S
113	E (12,4)	103	(52,41)	39*P*	(1, 0, −1, −1, 0, 0, −1)	10.3I + 20.6 M + 23S
149	E (13,1)	167	(32,133)	13*P*	(1, 0, 0, −1, −1)	7.6I + 15.3 M + 13.6S
1031	E (15,7)	1061	(217,808)	281*P*	(1, 0, 0, 1, 0, 0, −1, −1, −1)	13I + 26 M + 23S

Table 7.
The experimental results of the computational complexity for the scalar multiplications using the NAF method.

Some numerical results of the computational complexity to compute a scalar multiplication using the NAF method are given in **Table 7**.

11. The computational complexity for serial computing l-tuple of the scalar multiplications

11.1 The computational complexity of the serial GBR

On l-tuple of the scalar multiplications $\langle vP \rangle = \langle v_1P, v_2P, \dots, v_lP \rangle$, the computations of v_1P, v_2P, ..., v_lP without using the graphs or subgraphs can be done serially. So, the computational cost of these computations using the binary representations of v_1, v_2, ..., v_l is given by

$$\frac{t}{2}lA + tlD + 0.5tld. \tag{7}$$

In other words, the running time can be expressed in terms of field operations by

$$3tlS + 3tlM + 1.5tlI + 0.5tld. \tag{8}$$

Table 8 displays some small experimental results for computational complexities for serial computations of l-tuples $\langle vP \rangle$using the generalized binary method.

11.2 The computational complexity of the serial GNAF

The computational complexity for computing l-tuple of the scalar multiplications using GNAF representations in serial way is given by

$$lD + \frac{t}{3}lA + tlD = \frac{t}{3}lA + (t+1)lD. \tag{9}$$

Using the field operations, the formula in Eq. (9) can be rewritten by.

$$((t/3) + t + 1)lI + ((2/3)t + 2t + 2)lM + ((t/3) + 2t + 2)lS. \tag{10}$$

The computational complexity results for serial computations of l-tuples $\langle vP \rangle$ using the GNAF method are given in **Table 9**.

P	E (a,b)	n	Gen. pt. P	⟨vP⟩	Comp. complexity
101	E (10,2)	109	(68,14)	$\langle 93P, 25P, 66P \rangle$	63S + 63 M + 31.5I + 10.5d
61	E (4,1)	67	(24,14)	$\langle 23P, 19P, 12P \rangle$	45S + 45 M + 22.5I + 7.5d
113	E (12,4)	103	(52,41)	$\langle 39P, 21P \rangle$	36S + 36 M + 18I + 9d
149	E (13,1)	167	(32,133)	$\langle 13P, 5P \rangle$	24S + 24 M + 12I + 4d
1031	E (15,7)	1061	(217,808)	$\langle 281P, 91P, 63P, 55P \rangle$	108S + 108 M + 54I + 18d

Table 8.
The experimental results for computational complexities for serial computations of l*-tuples* $\langle vP \rangle$ *using the generalized binary method.*

P	E (a,b)	N	Gen. pt. P	$\langle vP \rangle$	Comp. complexity
101	E (10,2)	109	(68,14)	$\langle 93P, 25P, 66P \rangle$	34.8I + 69.9 M + 61.8S
61	E (4,1)	67	(24,14)	$\langle 23P, 19P, 12P \rangle$	27I + 54 M + 48S
113	E (12,4)	103	(52,41)	$\langle 39P, 21P \rangle$	20.6I + 41.2 M + 46S
149	E (13,1)	167	(32,133)	$\langle 13P, 5P \rangle$	15.2I + 30.6 M + 27.2S
1031	E (15,7)	1061	(217,808)	$\langle 281P, 91P, 63P, 55P \rangle$	52I + 104 M + 92S

Table 9.
The experimental results of the computational complexities for the serial computations of l-tuples $\langle vP \rangle$ using the GNAF.

12. The computational complexity of the graphic elliptic scalar multiplication methods

Suppose $\langle vP \rangle = \langle v_1P, v_2P, \dots, v_lP \rangle$ is an l-tuple of the scalar multiplications. The graphic computations of $v_1P, v_2P, \dots, v_lP$ can be done using the graphs or subgraphs in two ways. One of them is using the graphs directly to find the binary representations of the scalars $v_1, v_2, \dots, v_l,$ whereas another one uses the digraphs to represent these scalars. The computational costs of these computations can be discussed as follows.

12.1 The computational complexity of the graphic binary representation (GBR)

Using the graphs to compute l-tuple of the scalar multiplications costs

$$\frac{t}{2}lA + tlD. \tag{11}$$

In terms of field operations, the computational complexity of GBR can be expressed by

$$3tlS + 3tlM + 1.5tlI. \tag{12}$$

Table 10 displays some small experimental results for computational complexities for the graphic representations of l-tuples $\langle vP \rangle$ using the generalized binary method.

P	E (a,b)	n	Gen. pt. P	$\langle vP \rangle$	C_{GBR} using graphic representations
101	E (10,2)	109	(68,14)	$\langle 93P, 25P, 66P \rangle$	63S + 63 M + 31.5I
61	E (4,1)	67	(24,14)	$\langle 23P, 19P, 12P \rangle$	45S + 45 M + 22.5I
113	E (12,4)	103	(52,41)	$\langle 39P, 21P \rangle$	36S + 36 M + 18I
149	E (13,1)	167	(32,133)	$\langle 13P, 5P \rangle$	24S + 24 M + 12I
1031	E (15,7)	1061	(217,808)	$\langle 281P, 91P, 63P, 55P \rangle$	108S + 108 M + 54I

Table 10.
The experimental results for computational complexities for graphic computations of l-tuples $\langle vP \rangle$ using the generalized binary method.

P	E (a,b)	N	Gen. pt. P	$\langle vP \rangle$	C_{GBR} using graphic representations
101	E (10,2)	109	(68,14)	$\langle 93P, 25P, 66P \rangle$	32I + 64 M + 56S
61	E (4,1)	67	(24,14)	$\langle 23P, 19P, 12P \rangle$	24I + 48 M + 42S
113	E (12,4)	103	(52,41)	$\langle 39P, 21P \rangle$	18.6I + 37.3 M + 32.6S
149	E (13,1)	167	(32,133)	$\langle 13P, 5P \rangle$	13.3I + 26.6 M + 23.3S
1031	E (15,7)	1061	(217,808)	$\langle 281P, 91P, 63P, 55P \rangle$	48I + 96 M + 84S

Table 11.
The experimental results for computational complexities for graphic computations of l-tuples of $\langle vP \rangle$ using the GNAF method.

12.2 The computational complexity of the digraphic NAF

The computational complexity for computing l-tuple of the scalar multiplications using the digraphs is given by

$$\frac{t}{3}lA + tlD. \tag{13}$$

Eq. (13) can be rewritten using field operations by:

$$((\mathrm{t}/3) + \mathrm{t})l\mathrm{I} + ((2/3)\mathrm{t} + 2\,\mathrm{t})l\mathrm{M} + ((\mathrm{t}/3) + 2\,\mathrm{t})l\mathrm{S}. \tag{14}$$

Several experimental results for computational complexities for digraph representations of l-tuples $\langle vP \rangle$ are given in **Table 11**.

13. Computational complexity comparison on the serial and graphic computations of GBR and GNAF methods

This section discusses first the experimental results of the GBR method that uses serial computations to calculate l-tuple of the scalar multiplications and the GBR method that depends directly on using the graphs. Selecting the scalars $v_1, v_2, \ldots v_l$ from the interval [1. n − 1] to represent using the GBR method which needs the cost $0.5tld$, where t is the length of the string binary representation, l is the length of the tuple and d is a normal addition operation. The final computational cost as given in Eq. (8).

Whereas, the binary representing of the scalars $v_1, v_2, \ldots v_l$ can be taken directly from graphs or subgraphs without need to extra cost. This saves the $0.5tld$ operations to compute l-tuple of the scalar multiplications $\langle vP \rangle$. The total cost of the graphic GBR method has been determined previously in Eq. (12). The serial GBR and graphic GBR computational costs for several experimental results are given in **Table 12**. In this table, one can see the serial GBR method with various values of p is more costly compared to the graphic GBR method.

Also, the experimental results of the serial GNAF and graphic GNAF methods that are used to calculate l-tuple of the scalar multiplications are discussed in this section. Selecting the scalars $v_1, v_2, \ldots v_l$ from the interval [1. n − 1] to represent using the GNAF method which needs the $1l\mathrm{I} + 2l\mathrm{M} + 2l\mathrm{S}$ cost, l is the length of the tuple, M is a field multiplication, S is a field squaring, and I is a field inversion. So, the total computational cost as given in Eq. (10).

The graphic GNAF of the scalars $v_1, v_2, \ldots v_l$ can be taken directly from graphs. So it can save $1l\mathrm{I} + 2l\mathrm{M} + 2l\mathrm{S}$ operations for computing l-tuple of the scalar multiplications $\langle vP \rangle$. The total cost of the graphic GNAF method is determined

P	*E (a,b)*	*N*	Gen. pt. *P*	C_{GBR} using serial computations	C_{GBR} using graphs
101	*E* (10,2)	109	(68,14)	63S + 63 M + 31.5I + 10.5d	63S + 63 M + 31.5I
61	*E* (4,1)	67	(24,14)	45S + 45 M + 22.5I + 7.5d	45S + 45 M + 22.5I
113	E (12,4)	103	(52,41)	36S + 36 M + 18I + 9d	36S + 36 M + 18I
149	E (13,1)	167	(32,133)	24S + 24 M + 12I + 4d	24S + 24 M + 12I
1031	E (15,7)	1061	(217,808)	108S + 108 M + 54I + 18d	108S + 108 M + 54I

Table 12.
The computational costs of the serial GBR and graphic GBR with different values of p.

P	*E (a,b)*	*N*	Gen. pt. *P*	$Cost_{GNFA}$ using serial computations	$Cost_{GNFA}$ using graphs
101	*E* (10,2)	109	(68,14)	34.8I + 69.9 M + 61.8S	32I + 64 M + 56S
61	*E* (4,1)	67	(24,14)	27I + 54 M + 48S	24I + 48 M + 42S
113	E (12,4)	103	(52,41)	20.6I + 41.2 M + 46S	18.6I + 37.3 M + 32.6S
149	E (13,1)	167	(32,133)	15.2I + 30.6 M + 27.2S	13.3I + 26.6 M + 23.3S
1031	E (15,7)	1061	(217,808)	52I + 104 M + 92S	48I + 96 M + 84S

Table 13.
The computational costs of the serial GNAF and graphic GNAF with different values of p.

previously in Eq. (14). Several experimental results on the serial GNAF and graphic GNAF computational costs are given in **Table 13**. With various values of p as shown in **Table 13**, it can observe that the graphic GNAF method is less costly than the serial GNAF method.

14. Conclusions

The present chapter was concerned with presenting new graphic elliptic scalar multiplication algorithms for speeding up the computations of the scalar multiplication defined on elliptic curves over a prime field in different ways. These ways employed the undirected graphs and subgraphs to construct the binary representations of the scalars v in the scalar multiplications vP. Also, the sign digit representation of v has been obtained directly from using the digraphs or di-subgraphs. These representations are used to compute one scalar multiplication vP and l-tuple $<vP>$ of the scalar multiplications. The computational complexities of the proposed graphic elliptic scalar multiplication algorithms have been determined. The computational complexity comparison of the proposed algorithms and original ones is discussed based on the elliptic curve and field operations. The experiment results of the computational complexities show that the proposed algorithms are less costly for computing the scalar multiplication or l-tuple of the scalar multiplications than original algorithms which are dependent on the computations of the binary representations or NAF expansions. The new propositions with graphic representations speed up the computations on elliptic scalar multiplication algorithms. Also, it gives the generalized cases with the computations of the l-tuples $<vP>$ using (undirected or directed) graphs or subgraphs. This insight makes the working with graphic elliptic scalar multiplication algorithms more efficient in comparison with the serial original ones.

Author details

Ruma Kareem K. Ajeena
Mathematics Department, University of Babylon, Education College for Pure Sciences, Babil, Iraq

*Address all correspondence to: ruma.usm@gmail.com

References

[1] Hankerson D, Menezes AJ, Vanstone S. Guide to Elliptic Curve Cryptography. New York: Springer; 2004

[2] Hoffstein J et al. An Introduction to Mathematical Cryptography. Vol. 1. New York: Springer; 2008

[3] Oswald E. Introduction to Elliptic Curve Cryptography. Institute for Applied Information Processing and Communication, Graz University Technology; 2002

[4] Gross JL, Yellen J. Graph Theory and Its Applications. Chapman and Hall/ CRC; 2005

[5] Lopez J, Dahab R. An Overview of Elliptic Curve Cryptography. 2000

[6] Miller VS. Use of elliptic curves in cryptography. In: Conference on the heory and Application of Cryptographic Techniques. Berlin, Heidelberg: Springer; 1985

[7] Kapoor V, Abraham VS, Singh R. Elliptic curve cryptography. Ubiquity. 2008

[8] Karthikeyan E. Survey of elliptic curve scalar multiplication algorithms. International Journal of Advanced Networking and Applications. 2012;**4**(2)

[9] Brown M et al. Software implementation of the NIST elliptic curves over prime fields. In: Cryptographers' Track at the RSA Conference. Berlin, Heidelberg: Springer; 2001

[10] Dimitrov V, Imbert L, Mishra PK. Efficient and secure elliptic curve point multiplication using double-base chains. In: International Conference on the Theory and Application of Cryptology and Information Security. Berlin, Heidelberg: Springer; 2005

[11] Eisenträger K, Lauter K, Montgomery PL. Fast elliptic curve arithmetic and improved Weil pairing evaluation. In: Cryptographers' Track at the RSA Conference. Berlin, Heidelberg: Springer; 2003

[12] Ciet M et al. Trading inversions for multiplications in elliptic curve cryptography. Designs, Codes and Cryptography. 2006;**39**(2):189-206

[13] Mishra PK, Dimitrov V. Efficient quintuple formulas for elliptic curves and efficient scalar multiplication using multibase number representation. In: International Conference on Information Security. Berlin, Heidelberg: Springer; 2007

[14] Ajeena RKK, Kamarulhaili H. Point multiplication using integer sub-decomposition for elliptic curve cryptography. Applied Mathematics & Information Sciences. 2014;**8**(2):517

[15] Ajeena RKK, Kamarulhaili H. A hybrid approach for elliptic scalar multiplication. AIP Conference Proceedings. Vol. 1660. No. 1. AIP Publishing; 2015

[16] Ray SS. Graph theory with algorithms and its applications. In: Applied Science and Technology. Springer Science & Business Media; 2012

[17] Vasudev C. Graph Theory with Applications. New Age International; 2006

[18] Ajeena RKK, Kamarulhaili Hailiza. The computational complexity of elliptic curve integer sub- decomposition (ISD) method. I: AIP Conference Proceedings. Vol. 1605. No. 1. AIP; 2014

[19] Ajeena RKK. Integer Sub-decomposition (ISD) Method for Elliptic Curve Scalar Multiplication [Diss]. Universiti Sains Malaysia; 2015

Chapter 6

A Study of Bounded Variation Sequence Spaces

Vakeel Ahmad Khan, Hira Fatima and Mobeen Ahmad

Abstract

In the theory of classes of sequence, a wonderful application of Hahn-Banach extension theorem gave rise to the concept of Banach limit, i.e., the limit functional defined on c can be extended to the whole space l_∞ and this extended functional is called as the Banach limit. After that, in 1948 Lorentz used this concept of a week limit to introduce a new type of convergence, named as the almost convergence. Later on, Raimi generalized the concept of almost convergence known as $\sigma-$ convergence and the sequence space BV_σ was introduced and studied by Mursaleen. The main aim of this chapter is to study some new double sequence spaces of invariant means defined by ideal, modulus function and Orlicz function. Furthermore, we also study several properties relevant to topological structures and inclusion relations between these spaces.

Keywords: invariant mean, bounded variation, ideal, filter, I-convergence, Orlicz function, modulus function, paranorm

1. Introduction

The concept of convergence of a sequence of real numbers has been extended to statistical convergence independently by Fast [1] and Schoenberg [2]. There has been an effort to introduce several generalizations and variants of statistical convergence in different spaces. One such very important generalization of this notion was introduced by Kostyrko et al. [3] by using an ideal I of subsets of the set of natural numbers, which they called I-convergence. After that the idea of I-convergence for double sequence was introduced by Das et al. [4] in 2008.

Throughout a double sequence is defined by $x = (x_{ij})$ and we denote ${}_2\omega$ showing the space of all real or complex double sequences.

Let X be a nonempty set then a family $I \subset 2^X$ is said to be an **ideal** in X if $\emptyset \in I$, I is additive, i.e., for all $A, B \in I \Rightarrow A \cup B \in I$ and I is hereditary, i.e., for all $A \in I, B \subseteq A \Rightarrow B \in I$. A nonempty family of sets $\mathcal{F} \subset 2^X$ is said to be a **filter** on X if for all $A, B \in \mathcal{F}$ implies $A \cap B \in \mathcal{F}$ and for all $A \in \mathcal{F}$ with $A \subseteq B$ implies $B \in \mathcal{F}$. An ideal $I \subset 2^X$ is said to be **nontrivial** if $I \neq 2^X$, this non trivial ideal is said to be admissible if $I \supseteq \{\{x\} : x \in X\}$ and is said to be **maximal** if there cannot exist any nontrivial ideal $J \neq I$ containing I as a subset. For each ideal I there is a filter $\mathcal{F}(I)$ called as filter associate with ideal I, that is

$$\mathcal{F}(I) = \{K \subseteq X : K^c \in I\}, \quad \text{where} \quad K^c = X \backslash K. \tag{1}$$

A double sequence $x = (x_{ij}) \in {}_2\omega$ is said to be I-**convergent** [5–8] to a number L if for every $\epsilon>0$, we have $\{(i,j) \in \mathbb{N} \times \mathbb{N} : |x_{ij} - L| \geq \epsilon\} \in I$. In this case, we write $I - \lim x_{ij} = L$. A double sequence $x = (x_{ij}) \in {}_2\omega$ is said to be I-**Cauchy** if for every $\epsilon>0$ there exists numbers $m = m(\epsilon), n = n(\epsilon)$ such that $\{(i,j) \in \mathbb{N} \times \mathbb{N} : |x_{ij} - x_{mn}| \geq \epsilon\} \in I$.

A continuous linear functional ϕ on l_∞ is said to be an **invariant mean** [9, 10] or σ-mean if and only if:

1. $\phi(x) \geq 0$ where the sequence $x = (x_k)$ has $x_k \geq 0$ for all k,
2. $\phi(e) = 1$ where $e = \{1, 1, 1, 1, \ldots\}$,
3. $\phi(x_{\sigma(n)}) = \phi(x)$ for all $x \in l_\infty$,

where σ be an injective mapping of the set of the positive integers into itself having no finite orbits.

If $x = (x_k)$, write $Tx = (Tx_k) = (x_{\sigma(k)})$, so we have

$$V_\sigma = \left\{x = (x_k) : \lim_{m\to\infty} t_{m,k}(x) = L \ \text{ uniformly in } \ k, L = \sigma - \lim x\right\} \tag{2}$$

where $m \geq 0, k > 0$.

$$t_{m,k}(x) = \frac{x_k + x_{\sigma(k)} + \ldots + x_{\sigma^m(k)}}{m+1} \ \text{ and } \ t_{-1,k} = 0, \tag{3}$$

where $\sigma^m(k)$ denote the mth-iterate of $\sigma(k)$ at k. In this case σ is the translation mapping, that is, $\sigma(k) = k + 1$, $\sigma-$ mean is called a Banach limit [11] and V_σ, the set of bounded sequences of all whose invariant means are equal, is the set of almost convergent sequences. The special case of (3) in which $\sigma(k) = k + 1$ was given by Lorentz [12] and the general result can be proved in a similar way. It is familiar that a Banach limit extends the limit functional on c in the sense that

$$\phi(x) = \lim x, \quad \text{for all} \quad x \in \mathrm{c}. \tag{4}$$

Definition 1.1 A sequence $x \in l_\infty$ is of σ-**bounded variation** if and only if:

(i) $\sum|\phi_{m,k}(x)|$ converges uniformly in k,

(ii) $\lim_{m\to\infty} t_{m,k}(x)$, which must exist, should take the same value for all k.

We denote by BV_σ, the space of all sequences of σ-bounded variation:

$$BV_\sigma = \left\{x \in l_\infty : \sum_m |\phi_{m,k}(x)| < \infty, \ \text{ uniformly in } \ k\right\}.$$

is a Banach space normed by

$$\|x\| = \sup_k \sum_{m=0}^{\infty} |\phi_{m,k}(x)|. \tag{5}$$

A function $M : [0, \infty) \to [0, \infty)$ is said to be an **Orlicz function** [13, 14] if it satisfies the following conditions:

(i) M is continuous, convex and non-decreasing,

(ii) $M(0) = 0, M(x)>0$ and $M(x) \to \infty$ as $x \to \infty$.

Remark 1.1 If the convexity of an Orlicz function is replaced by $M(x + y) \leq M(x) + M(y)$, then this function is called **Modulus function** [15–17]. If M is an Orlicz function, then $M(\lambda X) \leq \lambda M(x)$ for all λ with $0 < \lambda < 1$. An Orlicz function M is said to satisfy Δ_2-condition for all values of u if there exists a constant $K > 0$ such that $M(Lu) \leq KLM(u)$ for all values of $L > 1$ [18].

Definition 1.2 A double sequence space X is said to be:

[i] **solid** or **normal** if $(x_{ij}) \in X$ implies that $(\alpha_{ij}x_{ij}) \in X$ for all sequence of scalars (α_{ij}) with $|\alpha_{ij}| < 1$ for all $(i,j) \in \mathbb{N} \times \mathbb{N}$.

[ii] **symmetric** if $(x_{\pi(i,j)}) \in X$ whenever $(x_{ij}) \in X$, where $\pi(i,j)$ is a permutation on $\mathbb{N} \times \mathbb{N}$.

[iii] **sequence algebra** if $\left(x_{ij}y_{ij}\right) \in E$ whenever (x_{ij}), $\left(y_{ij}\right) \in X$.

[iv] **convergence free** if $\left(y_{ij}\right) \in X$ whenever $(x_{ij}) \in X$ and $x_{ij} = 0$ implies $y_{ij} = 0$, for all $(i,j) \in \mathbb{N} \times \mathbb{N}$.

Definition 1.3 Let $K = \{(n_i, k_j) : (i,j) : n_1 < n_2 < n_3 < .... \text{ and } k_1 < k_2 < k_3 < ....\} \subseteq \mathbb{N} \times \mathbb{N}$ and X be a double sequence space. A K-step space of X is a sequence space

$$\lambda_k^E = \{(\alpha_{ij}x_{ij}) : (x_{ij}) \in X\}.$$

A canonical preimage of a sequence $\left(x_{n_ik_j}\right) \in X$ is a sequence $(b_{nk}) \in X$ defined as follows:

$$b_{nk} = \begin{cases} a_{nk}, & \text{for n, k} \in K \\ 0, & \text{otherwise.} \end{cases}$$

A sequence space X is said to be **monotone** if it contains the canonical preimages of all its step spaces.

The following subspaces $l(p), l_\infty(p), c(p)$ and $c_0(p)$ where $p = (p_k)$ is a sequence of positive real numbers. These subspaces were first introduced and discussed by Maddox [16]. The following inequalities will be used throughout the section. Let $p = \left(p_{ij}\right)$ be a double sequence of positive real numbers [19]. For any complex λ with $0 < p_{ij} \leq \sup_{ij} p_{ij} = G < \infty$, we have

$$|\lambda|^{p_{ij}} \leq \max\left(1, |\lambda|^G\right).$$

Let $D = \max(1, 2^{G-1})$ *and* $H = \max\left\{1, \sup_{ij} p_{ij}\right\}$, then for the sequences (a_{ij}) and (b_{ij}) in the complex plane, we have

$$|a_{ij} + b_{ij}|^{p_{ij}} \leq C(|a_{ij}|^{p_{ij}} + |b_{ij}|^{p_{ij}}).$$

2. Bounded variation sequence spaces defined by Orlicz function

In this section, we define and study the concepts of I-convergence for double sequences defined by Orlicz function and present some basic results on the above definitions [8, 20].

$$_2BV_\sigma^I(M) = \left\{(x_{ij}) \in {_2w} : I - \lim M\left(\frac{|\phi_{mnij}(x) - L|}{\rho}\right) = 0, \text{ for some } L \in \mathbb{C}, \rho > 0\right\} \tag{6}$$

$$_2\left({}_0BV_\sigma^I(M)\right) = \left\{(x_{ij}) \in {}_2w : I - \lim M\left(\frac{|\phi_{mnij}(x)|}{\rho}\right) = 0, \rho > 0\right\}, \tag{7}$$

$$_2\left({}_\infty BV_\sigma^I(M)\right) = \left\{(x_{ij}) \in {}_2w : \left\{(i,j) : \exists k > 0 \; s.t \; M\left(\frac{|\phi_{mnij}(x)|}{\rho}\right) \geq k\right\} \in I, \rho > 0\right\} \tag{8}$$

$$_2({}_\infty BV_\sigma(M)) = \left\{(x_{ij}) \in {}_2w : \sup M\left(\frac{|\phi_{mnij}(x)|}{\rho}\right) < \infty, \rho > 0\right\}. \tag{9}$$

Now, we read some theorems based on these sequence spaces. These theorems are of general importance as indispensable tools in various theoretical and practical problems.

Theorem 2.1 Let M_1, M_2 be two Orlicz functions with Δ_2 condition, then

(a) $\chi(M_2) \subseteq \chi(M_1M_2)$

(b) $\chi(M_1) \cap \chi(M_2) \subseteq \chi(M_1 + M_2)$ for $\chi = {}_2BV_\sigma^I, {}_2\left({}_0BV_\sigma^I\right)$.

Proof. (a) Let $x = (x_{ij}) \in {}_2\left({}_0BV_\sigma^I(M_2)\right)$ be an arbitrary element, so there exists $\rho > 0$ such that

$$I - \lim M_2\left(\frac{|\phi_{mnij}(x)|}{\rho}\right) = 0. \tag{10}$$

Let $\epsilon > 0$ and choose δ with $0 < \delta < 1$ such that $M_1(t) < \varepsilon$ for $0 < t \leq \delta$.

Write $y_{ij} = M_2\left(\frac{|\phi_{mnij}(x)|}{\rho}\right)$. Consider,

$$\lim_{ij} M_1\left(y_{ij}\right) = \lim_{y_{ij} \leq \delta,\, i,j \in \mathbb{N}} M_1\left(y_{ij}\right) + \lim_{y_{ij} > \delta,\, i,j \in \mathbb{N}} M_1\left(y_{ij}\right). \tag{11}$$

Now, since M_1 is an Orlicz function so we have $M_1(\lambda x) \leq \lambda M_1(x)$, $0 < \lambda < 1$. Therefore, we have

$$\lim_{y_{ij} \leq \delta,\, i,j \in \mathbb{N}} M_1\left(y_{ij}\right) \leq M_1(2) \lim_{y_{ij} \leq \delta,\, i,j \in \mathbb{N}} \left(y_{ij}\right). \tag{12}$$

For $y_{ij} > \delta$, we have $y_{ij} < \frac{y_{ij}}{\delta} < 1 + \frac{y_{ij}}{\delta}$. Now, since M_1 is non-decreasing and convex, it follows that,

$$M_1\left(y_{ij}\right) < M_1\left(1 + \frac{y_{ij}}{\delta}\right) < \frac{1}{2}M_1(2) + \frac{1}{2}M_1\left(\frac{2y_{ij}}{\delta}\right). \tag{13}$$

Since M_1 satisfies the Δ_2-condition, so we have

$$\begin{aligned} M_1\left(y_{ij}\right) &< \frac{1}{2}K\frac{y_{ij}}{\delta}M_1(2) + \frac{1}{2}KM_1\left(\frac{2y_{ij}}{\delta}\right) \\ &< \frac{1}{2}K\frac{y_{ij}}{\delta}M_1(2) + \frac{1}{2}K\frac{y_{ij}}{\delta}M_1(2) \\ &= K\frac{y_{ij}}{\delta}M_1(2). \end{aligned} \tag{14}$$

This implies that,

$$M_1\left(y_{ij}\right) < K\frac{y_{ij}}{\delta}M_1(2). \tag{15}$$

Hence, we have

$$\lim_{y_{ij}>\delta,\ i,j\in\mathbb{N}} M_1\left(y_{ij}\right) \le \max\left\{1, K\delta^{-1}M_1(2) \lim_{y_{ij}>\delta,\ i,j\in\mathbb{N}} \left(y_{ij}\right)\right\}. \tag{16}$$

Therefore from (12) and (16), we have

$$I - \lim_{ij} M_1\left(y_{ij}\right) = 0.$$

$$\Rightarrow I - \lim_{ij} M_1 M_2\left(\frac{|\phi_{mnij}(x)|}{\rho}\right) = 0.$$

This implies that $x = (x_{ij}) \in {}_2\left({}_0BV_\sigma^I(M_1M_2)\right)$. Hence $\chi(M_2) \subseteq \chi(M_1M_2)$ for $\chi = {}_2\left({}_0BV_\sigma^I\right)$. The other cases can be proved in similar way.

(b) Let $x = (x_{ij}) \in {}_2\left({}_0BV_\sigma^I(M_1)\right) \cap {}_2\left({}_0BV_\sigma^I(M_2)\right)$. Let $\epsilon>0$ be given. Then there exist $\rho>0$, such that

$$I - \lim_{ij} M_1\left(\frac{|\phi_{mnij}(x)|}{\rho}\right) = 0, \tag{17}$$

and

$$I - \lim_{ij} M_2\left(\frac{|\phi_{mnij}(x)|}{\rho}\right) = 0. \tag{18}$$

Therefore

$$I - \lim_{ij}\,(M_1 + M_2)\left(\frac{|\phi_{mnij}(x)|}{\rho}\right) = I - \lim_{ij} M_1\left(\frac{|\phi_{mnij}(x)|}{\rho}\right) + I - \lim_{ij} M_2\left(\frac{|\phi_{mnij}(x)|}{\rho}\right),$$

from Eqs. (17) and (18), we get

$$I - \lim_{ij}\,(M_1 + M_2)\left(\frac{|\phi_{mnij}(x)|}{\rho}\right) = 0.$$

so we have $x = (x_{ij}) \in {}_2\left({}_0BV_\sigma^I(M_1 + M_2)\right)$.

Hence, ${}_2\left({}_0BV_\sigma^I(M_1)\right) \cap {}_2\left({}_0BV_\sigma^I(M_2)\right) \subseteq {}_2\left({}_0BV_\sigma^I(M_1 + M_2)\right)$. For $\chi = {}_2BV_\sigma^I$ the inclusion are similar.

Corollary $\chi \subseteq \chi(M)$ *for* $\chi = {}_2\left(BV_\sigma^I\right)$ *and* ${}_2BV_\sigma^I$.

Proof. For this let $M(x) = x$, for all $x = (x_{ij}) \in X$. Let us suppose that $x = (x_{ij}) \in 2\left({}_0BV_\sigma^I\right)$. Then for any given $\epsilon>0$, we have

$$\left\{(i,j) : |\phi_{mnij}(x)| \ge \epsilon\right\} \in I.$$

Now let $A_1 = \left\{(i,j) : |\phi_{mnij}(x)| < \epsilon\right\} \in I$, be such that $A_1^c \in I$. Consider for $\rho > 0$,

$$M\left(\frac{|\phi_{mnij}(x)|}{\rho}\right) = \frac{|\phi_{mnij}(x)|}{\rho} < \frac{\epsilon}{\rho} < \epsilon.$$

This implies that $I - \lim M\left(\frac{|\phi_{mnij}(x)|}{\rho}\right) = 0$, which shows that $x = (x_{ij}) \in {}_2\left({}_0BV_\sigma^I(M)\right)$.

Hence, we have

$$_2({}_0BV_\sigma^I) \subseteq {}_2({}_0BV_\sigma^I(M)).$$
$$\Rightarrow \chi \subseteq \chi(M).$$

Using the definition of convergence free sequence space, let us give another theorem which will be of particular importance in our future work:

Theorem 2.2 The spaces ${}_2({}_0BV_\sigma^I(M))$ and ${}_2BV_\sigma^I(M)$ are not convergence free.

Example 2.1 To show this let $I = I_f$ and $M(x) = x$, for all $x = [0, \infty)$. Now consider the double sequence (x_{ij}), (y_{ij}) which defined as follows:

$$x_{ij} = \frac{1}{i+j} \text{ and } y_{ij} = i + j, \forall i, j \in \mathbb{N}.$$

Then we have (x_{ij}) belong to both ${}_2({}_0BV_\sigma^I(M))$ and ${}_2BV_\sigma^I(M)$, but (y_{ij}) does not belong to ${}_2({}_0BV_\sigma^I(M))$ and ${}_2BV_\sigma^I(M)$. Hence, the spaces ${}_2({}_0BV_\sigma^I(M))$ and ${}_2BV_\sigma^I(M)$ are not convergence free.

To gain a good understanding of these double sequence spaces and related concepts, let us finally look at this theorem on inclusions:

Theorem 2.3 Let M be an Orlicz function. Then

$${}_2({}_0BV_\sigma^I(M)) \subseteq {}_2BV_\sigma^I(M) \subseteq {}_2({}_\infty BV_\sigma^I(M)).$$

Proof. For this let us consider $x = (x_{ij}) \in {}_2({}_0BV_\sigma^I(M))$. It is obvious that it must belong to ${}_2BV_\sigma^I(M)$. Now consider

$$M\left(\frac{|\phi_{mnij}(x) - L|}{\rho}\right) \leq M\left(\frac{|\phi_{mnij}(x)|}{\rho}\right) + M\left(\frac{|L|}{\rho}\right).$$

Now taking the limit on both sides we get

$$I - \lim_{ij} M\left(\frac{|\phi_{mnij}(x) - L|}{\rho}\right) = 0.$$

Hence $x = (x_{ij}) \in {}_2BV_\sigma^I(M)$. Now it remains to show that

$${}_2(BV_\sigma^I(M)) \subseteq {}_2({}_\infty BV_\sigma^I(M)).$$

For this let us consider $x = (x_{ij}) \in {}_2BV_\sigma^I(M)$ this implies that there exist $\rho > 0$ s.t

$$I - \lim_{ij} M\left(\frac{|\phi_{mnij}(x) - L|}{\rho}\right) = 0.$$

Now consider,

$$M\left(\frac{|\phi_{mnij}(x)|}{\rho}\right) \leq M\left(\frac{|\phi_{mnij}(x) - L|}{\rho}\right) + M\left(\frac{|L|}{\rho}\right).$$

Now taking the supremum on both sides, we get

$$\sup_{ij} M\left(\frac{|\phi_{mnij}(x)|}{\rho}\right) < \infty.$$

Hence, $x = (x_{ij}) \in {}_2({}_\infty BV_\sigma^I(M))$. ■

3. Paranorm bounded variation sequence spaces

In this section we study double sequence spaces by using the double sequences of strictly positive real numbers $p = \left(p_{ij}\right)$ with the help of BV_σ space and an Orlicz function M. We study some of its properties and prove some inclusion relations related to these new spaces. For m, n ≥ 0, we have

$$_2BV_\sigma^I(M,p) = \{(x_{ij}) \in {_2\omega} : \left\{(i,j) : M\left(\frac{|\phi_{mnij}(x)-L|}{\rho}\right)^{p_{ij}} \geq \epsilon\right\} \in I; \text{ for some } L \in \mathbb{C}, \rho>0 \tag{19}$$

$$_2(_0BV_\sigma^I(M,p)) = \left\{(x_{ij}) \in {_2\omega} : \left\{(i,j) : M\left(\frac{|\phi_{mnij}(x)|}{\rho}\right)^{p_{ij}} \geq \epsilon\right\} \in I, \rho>0\right\}, \tag{20}$$

$$_2(_\infty BV_\sigma^I(M,p)) = \left\{(x_{ij}) \in {_2\omega} : \left\{(i,j) : \exists K>0 : M\left(\frac{|\phi_{mnij}(x)|}{\rho}\right)^{p_{ij}} \geq K\right\} \in I, \rho>0\right\} \tag{21}$$

$$_2l_\infty(M,p) = \left\{(x_{ij}) \in {_2\omega} : \sup M\left(\frac{|\phi_{mnij}(x)|}{\rho}\right)^{p_{ij}} < \infty, \rho > 0\right\}. \tag{22}$$

We also denote

$$_2M_{BV_\sigma}^I(M,p) = {_2BV_\sigma^I}(M,p) \cap {_2l_\infty}(M,p)$$

and

$$_2\left(_0M_{BV_\sigma}^I(M,p)\right) = {_2(_0BV_\sigma^I(M,p))} \cap {_2l_\infty}(M,p).$$

We can now state and proof the theorems based on these double sequence spaces which are as follows:

Theorem 3.1 Let $p = \left(p_{ij}\right) \in {_2l_\infty}$ then the classes of double sequence $_2M_{BV_\sigma}^I(M,p)$ and $_2\left(_0M_{BV_\sigma}^I(M,p)\right)$ are paranormed spaces, paranormed by

$$g(x_{ij}) = \inf_{i,j\geq 1}\left\{\rho^{\frac{p_{ij}}{H}} : \sup_{ij} M\left(\frac{|\phi_{mnij}(x)|}{\rho}\right)^{p_{ij}} \leq 1, \text{ for some} \rho > 0\right\}$$

where $H = \max\left\{1, \sup_{ij} p_{ij}\right\}$.

Proof. $(P1)$ It is clear that $g(x) = 0$ if and only if $x = 0$.

$(P2)$ $g(-x) = g(x)$ is obvious.

$(P3)$ Let $x = (x_{ij}), y = \left(y_{ij}\right) \in {_2M_{BV_\sigma}^I}(M,p)$. Now for $\rho_1, \rho_2 > 0$, we denote

$$A_1 = \left\{\rho_1 : \sup_{ij} M\left(\frac{|\phi_{mnij}(x)|}{\rho}\right)^{p_{ij}} \leq 1\right\} \tag{23}$$

$$A_2 = \left\{\rho_2 : \sup_{ij} M\left(\frac{|\phi_{mnij}(x)|}{\rho}\right)^{p_{ij}} \leq 1\right\} \tag{24}$$

Let us take $\rho_3 = \rho_1 + \rho_2$. Then by using the convexity of M, we have

$$M\left(\frac{|\phi_{mnij}(x+y)|}{\rho}\right) \leq \frac{\rho_1}{\rho_1+\rho_2}M\left(\frac{|\phi_{mnij}(x)|}{\rho_1}\right) + \frac{\rho_2}{\rho_1+\rho_2}M\left(\frac{|\phi_{mnij}(y)|}{\rho_2}\right)$$

which in terms give us

$$\sup_{ij} M\left(\frac{|\phi_{mnij}(x+y)|}{\rho}\right)^{p_{ij}} \leq 1$$

and

$$\begin{aligned} g\left(x_{ij}+y_{ij}\right) &= \inf\left\{(\rho_1+\rho_2)^{\frac{p_{ij}}{H}} : \rho_1 \in A_1, \rho_2 \in A_2\right\} \\ &\leq \inf\left\{(\rho_1)^{\frac{p_{ij}}{H}} : \rho_1 \in A_1\right\} + \inf\left\{(\rho_2)^{\frac{p_{ij}}{H}} : \rho_2 \in A_2\right\} \\ &= g(x_{ij}) + g\left(y_{ij}\right). \end{aligned}$$

Therefore $g(x+y) \leq g(x) + g(y)$.

(*P*4) Let (λ_{ij}) be a double sequence of scalars with $(\lambda_{ij}) \to \lambda \ (i,j \to \infty)$ and $(x_{ij}), L \in {}_2M^I_{BV_\sigma}(M,p)$ such that

$$x_{ij} \to L \ \ (i,j \to \infty),$$

in the sense that

$$g(x_{ij} - L) \to 0 \ \ (i,j \to \infty).$$

Then, since the inequality

$$g(x_{ij}) \leq g(x_{ij} - L) + g(L)$$

holds by subadditivity of g, the sequence $g(x_{ij})$ is bounded.

Therefore,

$$\begin{aligned} g\left[(\lambda_{ij}x_{ij} - \lambda L)\right] &= g\left[(\lambda_{ij}x_{ij} - \lambda x_{ij} + \lambda x_{ij} - \lambda L)\right] \\ &= g\left[(\lambda_{ij} - \lambda)x_{ij} + \lambda(x_{ij} - L)\right] \\ &\leq g\left[(\lambda_{ij} - \lambda)x_{ij}\right] + g\left[\lambda(x_{ij} - L)\right] \\ &\leq \left|(\lambda_{ij} - \lambda)\right|^{\frac{p_{ij}}{M}} g(x_{ij}) + |\lambda|^{\frac{p_{ij}}{M}} g(x_{ij} - L) \to 0 \end{aligned}$$

as $(i,j \to \infty)$. That implies that the scalar multiplication is continuous. Hence ${}_2M^I_{BV_\sigma}(M,p)$ is a paranormed space. For another space ${}_2\left({}_0M^I_{BV_\sigma}(M,p)\right)$, the result is similar.

We shall see about the separability of these new defined double sequence spaces in the next theorem.

Theorem 3.2 The spaces ${}_2M^I_{BV_\sigma}(M,p)$ and ${}_2\left({}_0M^I_{BV_\sigma}(M,p)\right)$ are not separable.

Example 3.1 By counter example, we prove the above result for the space ${}_2M^I_{BV_\sigma}(M,p)$.

Let A be an infinite subset of increasing natural numbers, i.e., $A \subseteq \mathbb{N} \times \mathbb{N}$ such that $A \in I$.

Let

$$p_{ij} = \begin{cases} 1, & \text{if} (i,j) \in A \\ 2, & \text{otherwise}. \end{cases}$$

Let $P_0 = \{(x_{ij}) : x_{ij} = 0 \text{ or } 1, \text{for } i,j \in M \text{ and } x_{ij} = 0, \text{otherwise}\}$.

Since A is infinite, so P_0 is uncountable. Consider the class of open balls

$$B_1 = \left\{B\left(z, \frac{1}{2}\right) : z \in P_0\right\}.$$

Let C_1 be an open cover of ${}_2M^I_{BV_\sigma}(M,p)$ containing B_1.

Since B_1 is uncountable, so C_1 cannot be reduced to a countable subcover for ${}_2M^I_{BV_\sigma}(M,p)$. Thus ${}_2M^I_{BV_\sigma}(M,p)$ is not separable.

We shall now introduce a theorem which improves our work.

Theorem 3.3 Let (p_{ij}) and (q_{ij}) be two double sequences of positive real numbers. Then ${}_2\left({}_0M^I_{BV_\sigma}(M,p)\right) \supseteq {}_2\left({}_0M^I_{BV_\sigma}(M,q)\right)$ if and only if $\lim_{i,j\in K} \inf \frac{p_{ij}}{q_{ij}} > 0$, where $K^c \subseteq \mathbb{N} \times \mathbb{N}$ such that $K \in I$.

Proof. Let $\lim_{i,j\in K} \inf \frac{p_{ij}}{q_{ij}} > 0$ and $(x_{ij}) \in {}_2\left({}_0M^I_{BV_\sigma}(M,q)\right)$. Then, there exists $\beta > 0$ such that $p_{ij} > \beta q_{ij}$ for sufficiently large $(i,j) \in K$.

Since $(x_{ij}) \in {}_2\left({}_0M^I_{BV_\sigma}(M,q)\right)$. For a given $\epsilon > 0$, there exist $\rho > 0$ such that

$$B_0 = \left\{(i,j) \in \mathbb{N} \times \mathbb{N} : M\left(\frac{|\phi_{mnij}(x)|}{\rho}\right)^{q_{ij}} \geq \epsilon\right\} \in I.$$

Let $G_0 = K^c \cup B_0$. Then for all sufficiently large $(i,j) \in G_0$.

$$\left\{(i,j) : M\left(\frac{|\phi_{mnij}(x)|}{\rho}\right)^{p_{ij}} \geq \epsilon\right\} \subseteq \left\{(i,j) : M\left(\frac{|\phi_{mnij}(x)|}{\rho}\right)^{\beta q_{ij}} \geq \epsilon\right\} \in I.$$

Therefore, $(x_{ij}) \in {}_2\left({}_0M^I_{BV_\sigma}(M,p)\right)$. The converse part of the result follows obviously.

Remark 3.1 Let (p_{ij}) and (q_{ij}) be two double sequences of positive real numbers. Then ${}_2\left({}_0M^I_{BV_\sigma}(M,q)\right) \supseteq {}_2\left({}_0M^I_{BV_\sigma}(M,p)\right)$ if and only if $\lim_{i,j\in K} \inf \frac{q_{ij}}{p_{ij}} > 0$ and ${}_2\left({}_0M^I_{BV_\sigma}(M,q)\right) = {}_2\left({}_0M^I_{BV_\sigma}(M,p)\right)$ if and only if $\lim_{i,j\in K} \inf \frac{p_{ij}}{q_{ij}} > 0$ and $\lim_{i,j\in K} \inf \frac{q_{ij}}{p_{ij}} > 0$, where $K^c \subseteq \mathbb{N} \times \mathbb{N}$ such that $K \in I$.

Theorem 3.4 The set ${}_2M^I_{BV_\sigma}(M,p)$ is closed subspace of ${}_2l_\infty(M,p)$.

Proof. Let $\left(x^{(pq)}_{ij}\right)$ be a Cauchy double sequence in ${}_2M^I_{BV_\sigma}(M,p)$ such that $x^{(pq)} \to x$. We show that $x \in {}_2M^I_{BV_\sigma}(M,p)$. Since, $\left(x^{(pq)}_{ij}\right) \in {}_2M^I_{BV_\sigma}(M,p)$, then there exists a_{pq}, and $\rho > 0$ such that

$$\left\{(i,j) : M\left(\frac{|\phi_{mnij}(x^{pq}) - a_{pq}|}{\rho}\right)^{p_{ij}} \geq \epsilon\right\} \in I.$$

We need to show that

(1) (a_{pq}) converges to a.

(2) If $U = \left\{(i,j) : M\left(\frac{|\phi_{mnij}(x^{pq}) - a|}{\rho}\right)^{p_{ij}} < \epsilon\right\}$, then $U^c \in I$.

Since $\left(x_{ij}^{(pq)}\right)$ be a Cauchy double sequence in ${}_2M^I_{BV_\sigma}(M,p)$ then for a given $\epsilon>0$ there exists $k_0\in\mathbb{N}$ such that

$$\sup_{ij} M\left(\frac{|\phi_{mnij}(x^{pq})-\phi_{mnij}(x^{rs})|}{\rho}\right)^{p_{ij}} < \frac{\epsilon}{3}, \text{ for all } p,q,r,s\geq k_0.$$

For a given $\epsilon>0$, we have

$$B_{pqrs}=\left\{(i,j):M\left(\frac{|\phi_{mnij}(x^{pq})-\phi_{mnij}(x^{rs})|}{\rho}\right)^{p_{ij}}<\left(\frac{\epsilon}{3}\right)^M\right\},$$
$$B_{pq}=\left\{(i,j):M\left(\frac{|\phi_{mnij}(x^{pq})-a_{pq}|}{\rho}\right)^{p_{ij}}<\left(\frac{\epsilon}{3}\right)^M\right\},$$
$$B_{rs}=\left\{(i,j):M\left(\frac{|\phi_{mnij}(x^{rs})-a_{rs}|}{\rho}\right)^{p_{ij}}<\left(\frac{\epsilon}{3}\right)^M\right\}.$$

Then $B^c_{pqrs}, B^c_{pq}, B^c_{rs}\in I$. Let $B^c=B^c_{pqrs}\cap B^c_{pq}\cap B^c_{rs}$,
where $B=\left\{(i,j):M\left(\frac{|a_{pq}-a_{rs}|}{\rho}\right)^{p_{ij}}<\epsilon\right\}$, then $B^c\in I$. We choose $k_0\in B^c$, then for each $p,q,r,s\geq k_0$, we have

$$\left\{(i,j):M\left(\frac{|a_{pq}-a_{rs}|}{\rho}\right)^{p_{ij}}<\epsilon\right\}\supseteq\left[\left\{i,j\in\mathbb{N}:M\left(\frac{|\phi_{mnij}(x^{pq})-\phi_{mnij}(x^{rs})|}{\rho}\right)^{p_{ij}}<\left(\frac{\epsilon}{3}\right)^M\right\}\right.$$
$$\cap\left\{(i,j):M\left(\frac{|\phi_{mnij}(x^{pq})-a_{pq}|}{\rho}\right)^{p_{ij}}<\left(\frac{\epsilon}{3}\right)^M\right\}$$
$$\left.\cap\left\{(i,j):M\left(\frac{|\phi_{mnij}(x^{rs})-a_{rs}|}{\rho}\right)^{p_{ij}}<\left(\frac{\epsilon}{3}\right)^M\right\}\right].$$

Then (a_{pq}) is a Cauchy double sequence in $\mathbb{C}$. So, there exists a scalar $a\in\mathbb{C}$ such that $(a_{pq})\to a$, *as* $p,q\to\infty$.

(2) For the next step, let $0<\delta<1$ be given. Then, we show that if

$$U=\left\{(i,j):M\left(\frac{|\phi_{mnij}(x^{pq})-a|}{\rho}\right)^{p_{ij}}\leq\delta\right\}$$

then $U^c\in I$. Since $x^{(pq)}\to x$, then there exists $p_0,q_0\in\mathbb{N}$ such that,

$$P=\left\{(i,j):M\left(\frac{|\phi_{mnij}(x^{p_0q_0})-\phi_{mnij}(x)|}{\rho}\right)^{p_{ij}}<\left(\frac{\delta}{3D}\right)^H\right\}\tag{25}$$

where $D=\max\{1,2^{G-1}\}$, $G=\sup_{ij}p_{ij}\geq 0$ and $H=\max\left\{1,\sup_{ij}p_{ij}\right\}$ implies $P^c\in I$. The number (p_0,q_0) can be so chosen that together with (25), we have

$$Q=\left\{(i,j):M\left(\frac{|a_{p_0q_0}-a|}{\rho}\right)^{p_{ij}}<\left(\frac{\delta}{3D}\right)^H\right\}$$

such that $Q^c\in I$. Since $\left(x_{ij}^{(pq)}\right)\in {}_2M^I_{BV_\sigma}(M,p)$.

We have

$$\left\{(i,j):M\left(\frac{|\phi_{mnij}(x^{p_0q_0})-a_{p_0q_0}|}{\rho}\right)^{p_{ij}}\geq\delta\right\}\in I.$$

Then we have a subset $S \subseteq \mathbb{N} \times \mathbb{N}$ such that $S^c \in I$, where

$$S = \left\{ (i,j) : M\left(\frac{|\phi_{mnij}(x^{p_0q_0}) - a_{p_0q_0}|}{\rho}\right)^{p_{ij}} < \left(\frac{\delta}{3D}\right)^H \right\}.$$

Let $U^c = P^c \cup Q^c \cup S^c$, where

$$U = \left\{ (i,j) : M\left(\frac{|\phi_{mnij}(x) - a|}{\rho}\right)^{p_{ij}} < \delta \right\}$$

Therefore, for $(i,j) \in U^c$, we have

$$\begin{aligned} &\left\{ (i,j) : M\left(\frac{|\phi_{mnij}(x)-a|}{\rho}\right)^{p_{ij}} < \delta \right\} \\ &\supseteq \left[\left\{ (i,j) : M\left(\frac{|\phi_{mnij}(x^{p_0q_0})-\phi_{mnij}(x)|}{\rho}\right)^{p_{ij}} < \left(\frac{\delta}{3D}\right)^H \right\} \right. \\ &\cap \left\{ (i,j) : M\left(\frac{|a_{p_0q_0} - a|}{\rho}\right)^{p_{ij}} < \left(\frac{\delta}{3D}\right)^M \right\} \\ &\left. \cap \left\{ (i,j) : M\left(\frac{|\phi_{mnij}(x^{p_0q_0}) - a_{p_0q_0}|}{\rho}\right)^{p_{ij}} < \left(\frac{\delta}{3D}\right)^H \right\} \right]. \end{aligned}$$

Hence the result ${}_2M^I_{BV_\sigma}(M,p) \subset {}_2l_\infty(M,p)$ follows.

Since the inclusions ${}_2M^I_{BV_\sigma}(M,p) \subset {}_2l_\infty(M,p)$ and ${}_2\left({}_0M^I_{BV_\sigma}(M,p)\right) \subset {}_2l_\infty(M,p)$ are strict so in view of Theorem (3.3), we have the following result.

The above theorem is interesting and itself will have various applications in our future work.

4. Bounded variation sequence spaces defined by modulus function

In this section, we study some new double sequence spaces of invariant means defined by ideal and modulus function. Furthermore, we also study several properties relevant to topological structures and inclusion relations between these spaces. The following classes of double sequence spaces are as follows:

$${}_2BV^I_\sigma(f) = \left\{ (x_{ij}) \in {}_2\omega : \left\{ (i,j) : \sum_{m,\,n=0}^{\infty} f\left(|\phi_{mnij}(x) - L|\right) \geq \epsilon \right\} \in I; \text{for some } L \in \mathbb{C} \right\}; \tag{26}$$

$${}_2\left({}_0BV^I_\sigma(f)\right) = \left\{ (x_{ij}) \in {}_2\omega : \left\{ (i,j) : \sum_{m,\,n=0}^{\infty} f\left(|\phi_{mnij}(x)|\right) \geq \epsilon \right\} \in I \right\}; \tag{27}$$

$${}_2\left({}_\infty BV^I_\sigma(f)\right) = \left\{ (x_{ij}) \in {}_2\omega : \left\{ (i,j) : \exists K>0 : \sum_{m,\,n=0}^{\infty} f\left(|\phi_{mnij}(x)|\right) \geq K \right\} \in I \right\}; \tag{28}$$

$${}_2\left({}_\infty BV_\sigma(f)\right) = \left\{ (x_{ij}) \in {}_2\omega : \sup_{i,j} \sum_{m,\,n=0}^{\infty} f\left(|\phi_{mnij}(x)|\right) < \infty \right\}. \tag{29}$$

We also denote

$$_2M^I_{BV_\sigma}(f) = {}_2BV^I_\sigma(f) \cap {}_2({}_\infty BV_\sigma(f))$$

and

$$_2\left({}_0M^I_{BV_\sigma}(f)\right) = {}_2\left({}_0BV^I_\sigma(f)\right) \cap {}_2({}_\infty BV_\sigma(f)).$$

We shall now consider important theorems of these double sequence spaces by using modulus function.

Theorem 4.1 *For any modulus function f, the classes of double sequence* $_2({}_0BV^I_\sigma(f))$, $_2BV^I_\sigma(f)$, $_2\left({}_0M^I_{BV_\sigma}(f)\right)$ *and* $_2M^I_{BV_\sigma}(f)$ *are linear spaces.*

Proof. Suppose $x = (x_{ij})$ and $y = (y_{ij}) \in {}_2BV^I_\sigma(f)$ be any two arbitrary elements. Let α, β are scalars. Now, since $(x_{ij}), (y_{ij}) \in {}_2BV^I_\sigma(f)$. Then this implies that there exists some positive numbers $L_1, L_2 \in \mathbb{C}$ and such that the sets

$$A_1 = \left\{(i,j) : \sum_{m,\,n=0}^{\infty} f\left(|\phi_{mnij}(x) - L_1|\right) \geq \frac{\epsilon}{2}\right\} \in I, \tag{30}$$

$$A_2 = \left\{(i,j) : \sum_{m,\,n=0}^{\infty} f\left(|\phi_{mnij}(y) - L_2|\right) \geq \frac{\epsilon}{2}\right\} \in I. \tag{31}$$

Now, assume

$$B_1 = \left\{(i,j) : \sum_{m,\,n=0}^{\infty} f\left(|\phi_{mnij}(x) - L_1|\right) < \frac{\epsilon}{2}\right\} \in \mathcal{F}(I), \tag{32}$$

$$B_2 = \left\{(i,j) : \sum_{m,\,n=0}^{\infty} f\left(|\phi_{mnij}(y) - L_2|\right) < \frac{\epsilon}{2}\right\} \in \mathcal{F}(I) \tag{33}$$

be such that $B_1^c, B_2^c \in I$. Since f is a modulus function, we have

$$\begin{aligned}
&\sum_{m,\,n=0}^{\infty} f\left(|\phi_{mnij}(\alpha x + \beta y) - (\alpha L_1 + \beta L_2)|\right)\\
&= \sum_{m,\,n=0}^{\infty} f\left(\left|\left(\alpha\phi_{mnij}(x) + \beta\phi_{mnij}(y)\right) - (\alpha L_1 + \beta L_2)\right|\right)\\
&= \sum_{m,\,n=0}^{\infty} f\left(\left|\alpha\left(\phi_{mnij}(x) - L_1\right) + \beta\left(\phi_{mnij}(y) - L_2\right)\right|\right)\\
&\leq \sum_{m,\,n=0}^{\infty} f\left(|\alpha||\phi_{mnij}(x) - L_1|\right) + \sum_{m,\,n=0}^{\infty} f\left(|\beta||\phi_{mnij}(y) - L_2|\right)\\
&\leq \sum_{m,\,n=0}^{\infty} f\left(|\phi_{mnij}(x) - L_1|\right) + \sum_{m,\,n=0}^{\infty} f\left(|\phi_{mnij}(y) - L_2|\right)\\
&\leq \frac{\epsilon}{2} + \frac{\epsilon}{2} = \epsilon
\end{aligned}$$

This implies that $\left\{(i,j) : \sum_{m,\,n=0}^{\infty} f\left(|\phi_{mnij}(\alpha x + \beta y) - (\alpha L_1 + \beta L_2)|\right) \geq \epsilon\right\} \in I$. Thus $\alpha(x_{ij}) + \beta(y_{ij}) \in {}_2BV^I_\sigma(f)$. As (x_{ij}) and (y_{ij}) are two arbitrary element then

$\alpha(x_{ij}) + \beta(y_{ij}) \in {}_2BV_\sigma^I(f)$ for all (x_{ij}), $(y_{ij}) \in {}_2BV_\sigma^I(f)$ and for all scalars α, β. Hence ${}_2BV_\sigma^I(f)$ is linear space. The proof for other spaces will follow similarly. ■

We may go a step further and define another theorem on ideal convergence which basically depends upon the set in the filter associated with the same ideal.

Theorem 4.2 A sequence $x = (x_{ij}) \in {}_2M_{BV_\sigma}^I(f)$ I-convergent if and only if for every $\epsilon>0$, there exists $M_\varepsilon, N_\varepsilon \in \mathbb{N}$ such that

$$\left\{(i,j) : \sum_{m,\,n=0}^{\infty} f\left(|\phi_{mnij}(x_{ij}) - \phi_{mnij}(x_{M_\epsilon,N_\epsilon})|\right) < \epsilon\right\} \in \mathcal{F}(I).$$

Proof. Let $x = (x_{ij}) \in {}_2M_{BV_\sigma}^I(f)$. Suppose $I - \lim x = L$. Then, the set

$$B_\epsilon = \left\{(i,j) : \sum_{m,\,n=0}^{\infty} f\left(|\phi_{mnij}(x_{ij}) - L|\right) < \frac{\epsilon}{2}\right\} \in F(I), \quad \text{for all} \quad \epsilon>0.$$

Fix $M_\varepsilon, N_\varepsilon \in B_\varepsilon$. Then we have

$$\begin{aligned}\sum_{m,\,n=0}^{\infty} f\left(|\phi_{mnij}(x_{ij}) - \phi_{mnij}(x_{M_\epsilon,N_\epsilon})|\right) &\le \sum_{m,\,n=0}^{\infty} f\left(|\phi_{mnij}(x_{M_\epsilon,N_\epsilon}) - L|\right)\\ &+ \sum_{m,\,n=0}^{\infty} f\left(|L - \phi_{mnij}(x_{ij})|\right)\\ &< \frac{\epsilon}{2} + \frac{\epsilon}{2} = \epsilon\end{aligned}$$

which holds for all $(i,j) \in B_\epsilon$.
Hence

$$\left\{(i,j) : \sum_{m,\,n=0}^{\infty} f\left(|\phi_{mnij}(x_{ij}) - \phi_{mnij}(x_{M_\epsilon,N_\epsilon})|\right) < \epsilon\right\} \in \mathcal{F}(I).$$

Conversely, suppose that

$$\left\{(i,j) : \sum_{m,\,n=0}^{\infty} f\left(|\phi_{mnij}(x_{ij}) - \phi_{mnij}(x_{M_\epsilon,N_\epsilon})|\right) < \epsilon\right\} \in \mathcal{F}(I).$$

Then, being f a modulus function and by using basic triangular inequality, we have

$$\left\{(i,j) : \left|\sum_{m,\,n=0}^{\infty} f\left(|\phi_{mnij}(x_{ij})|\right) - \sum_{m,\,n=0}^{\infty} f\left(|\phi_{mnij}(x_{M_\epsilon,N_\epsilon})|\right)\right| < \epsilon\right\} \in F(I), \text{ for all } \epsilon>0.$$

Then, the set

$$C_\varepsilon = \left\{(i,j) : \sum_{m,\,n=0}^{\infty} f\left(|\phi_{mnij}(x_{ij})|\right) \in \left[\sum_{m,\,n=0}^{\infty} f\left(|\phi_{mnij}(x_{M_\epsilon,N_\epsilon})|\right) - \epsilon, \sum_{m,\,n=0}^{\infty} f\left(|\phi_{mnij}(x_{M_\epsilon,N_\epsilon})|\right) + \epsilon\right]\right\} \in \mathcal{F}(I).$$

Let $J_\epsilon = \left[\sum_{m,\,n=0}^{\infty} f\left(|\phi_{mnij}(x_{M_\epsilon,N_\epsilon})|\right) - \epsilon, \sum_{m,\,n=0}^{\infty} f\left(|\phi_{mnij}(x_{M_\epsilon,N_\epsilon})|\right) + \epsilon\right]$.

If we fix $\epsilon>0$ then, we have $C_\epsilon \in F(I)$ as well as $C_{\frac{\epsilon}{2}} \in \mathcal{F}(I)$.
Hence $C_\epsilon \cap C_{\frac{\epsilon}{2}} \in \mathcal{F}(I)$. This implies that

$$J = J_\epsilon \cap J_{\frac{\epsilon}{2}} \neq \phi.$$

That is

$$\left\{(i,j) : \sum_{m,\, n=0}^{\infty} f\left(|\phi_{mnij}(x_{ij})|\right) \in J\right\} \in \mathcal{F}(I).$$

This shows that

$$diam\, J \leq diam\, J_\epsilon$$

where the $diam\, J$ denotes the length of interval J. In this way, by induction we get the sequence of closed intervals

$$J_\epsilon = I_0 \supseteq I_1 \supseteq I_2 \supseteq \ldots \supseteq I_k \supseteq \ldots$$

with the property that $diam\, I_k \leq \frac{1}{2} diam\, I_{k-1}$ for $(k = 2, 3, 4, \ldots)$ and $\left\{(i,j) : \sum_{m,\, n=0}^{\infty} f\left(|\phi_{mnij}(x_{ij})|\right) \in I_k\right\} \in F(I)$ for $(k = 1, 2, 3, 4, \ldots)$.
Then there exists a $\xi \in \cap I_k$ where $k \in \mathbb{N}$ such that

$$\xi = I - \lim_{i,j} \sum_{m,\, n=0}^{\infty} f\left(|\phi_{mnij}(x_{ij})|\right),$$

showing that $x = (x_{ij}) \in {}_2M^I_{BV_\sigma}(f)$ is I-convergent. Hence the result holds.

As the reader knows about solid and monotone sequence space now turn to theorem on solid and monotone double sequence spaces of invariant mean defined by ideal and modulus function.

Theorem 4.3 For any modulus function f, the spaces ${}_2({}_0BV^I_\sigma(f))$ and ${}_2\left({}_0M^I_{BV_\sigma}(f)\right)$ are solid and monotone.

Proof. We consider ${}_2({}_0BV^I_\sigma(f))$ and for ${}_2\left({}_0M^I_{BV_\sigma}(f)\right)$ the proof shall be similar.

Let $x = (x_{ij}) \in {}_2({}_0BV^I_\sigma(f))$ be an arbitrary element, then the set

$$\left\{(i,j) : \sum_{m,\, n=0}^{\infty} f\left(|\phi_{mnij}(x)|\right) \geq \epsilon\right\} \in I. \tag{34}$$

Let (α_{ij}) be a sequence of scalars with $|\alpha_{ij}| \leq 1$ for all $i, j \in \mathbb{N}$.

Now, since f is a modulus function. Then the result follows from (2.18) and the inequality

$$f\left(|\alpha_{ij}\phi_{mnij}(x)|\right) \leq |\alpha_{ij}| f\left(|\phi_{mnij}(x)|\right) \leq f\left(|\phi_{mnij}(x)|\right).$$

Therefore,

$$\left\{(i,j) : \sum_{m,\, n=0}^{\infty} f\left(|\alpha_{ij}\phi_{mnij}(x)|\right) \geq \epsilon\right\} \subseteq \left\{(i,j) : \sum_{m,\, n=0}^{\infty} f\left(|\phi_{mnij}(x)|\right) \geq \epsilon\right\} \in I$$

implies that

$$\left\{(i,j) : \sum_{m,\,n=0}^{\infty} f\left(|\alpha_{ij}\phi_{mnij}(x)|\right) \geq \epsilon\right\} \in I.$$

Thus we have $(\alpha_{ij}x_{ij}) \in {}_2({}_0BV_\sigma^I(f))$. Hence ${}_2({}_0BV_\sigma^I(f))$ is solid. Therefore ${}_2({}_0BV_\sigma^I(f))$ is monotone. Since every solid sequence space is monotone.

Remark 4.1 The space ${}_2BV_\sigma^I(f)$ and ${}_2\left(M_{BV_\sigma^I}(f)\right)$ are neither solid nor monotone in general.

Example 4.1 Here we give counter example for establishment of this result. Let $X{=}_2BV_\sigma^I$ and ${}_2\left(M_{BV_\sigma^I}\right)$. Let us consider $I = I_f$ and $f(x) = x$, for all $x = (x_{ij})$ and $x_{ij} \in [0, \infty)$. Consider, the K-step space $X_K(f)$ of $X(f)$ defined as follows:

Let $x = (x_{ij}) \in X(f)$ and $y = (y_{ij}) \in X_K(f)$ be such that

$$y_{ij} = \begin{cases} x_{ij}, & \text{if } i,\ j \text{ are even} \\ 0, & \text{otherwise.} \end{cases}$$

Consider the sequence (x_{ij}) defined by $(x_{ij}) = 1$ for all $i, j \in \mathbb{N}$.

Then, $x = (x_{ij}) \in {}_2BV_\sigma^I(f)$ and ${}_2M_{BV_\sigma^I}(f)$, but K-step space preimage does not belong to $BV_\sigma^I(f)$ and ${}_2M_{BV_\sigma}^I(f)$. Thus, ${}_2BV_\sigma^I(f)$ and ${}_2M_{BV_\sigma}^I(f)$ are not monotone and hence they are not solid.

After discussing about solid and monotone sequence space now we come to the concept of sequence algebra which will help to understand our further work.

Theorem 4.4 For any modulus function f, the spaces ${}_2({}_0BV_\sigma^I(f))$ and ${}_2BV_\sigma^I(f)$ are sequence algebra.

Proof. Let $x = (x_{ij}), y = (y_{ij}) \in {}_2({}_0BV_\sigma^I(f))$ be any two arbitrary elements.

Then, the sets

$$\left\{(i,j) : \sum_{m,\,n=0}^{\infty} f\left(|\phi_{mnij}(x)|\right) \geq \epsilon\right\} \in I$$

and

$$\left\{(i,j) : \sum_{m,\,n=0}^{\infty} f\left(|\phi_{mnij}(y)|\right) \geq \epsilon\right\} \in I.$$

Therefore,

$$\left\{(i,j) : \sum_{m,\,n=0}^{\infty} f\left(|\phi_{mnij}(x).\phi_{mnij}(y)|\right) \geq \epsilon\right\} \in I.$$

Thus, we have $(x_{ij}).(y_{ij}) \in {}_2({}_0BV_\sigma^I(f))$. Hence ${}_2({}_0BV_\sigma^I(f))$ is sequence algebra. And for ${}_2BV_\sigma^I(f)$ the result can be proved similarly.

Remark 4.2 If I is not maximal and $I \neq I_f$ then the spaces ${}_2BV_\sigma^I(f)$ and ${}_2({}_0BV_\sigma^I(f))$ are not symmetric.

Example 4.2 Let $A \in I$ be an infinite set and $f(x) = x$ for all $x = (x_{ij})$ and $x_{ij} \in [0, \infty)$. If

$$x_{ij} = \begin{cases} 1, & \text{if } (i,j) \in A \\ 0, & \text{otherwise} \end{cases}$$

Then, it is clearly seen that $(x_{ij}) \in {}_2({}_0BV_\sigma^I(f)) \subset {}_2BV_\sigma^I(f)$.

Let $K \subseteq \mathbb{N} \times \mathbb{N}$ be such that $K \notin I$ and $K^c \notin I$. Let $\phi : K \to A$ and $\psi : K^c \to A^c$ be a bijective maps (as all four sets are infinite). Then, the mapping $\pi : \mathbb{N} \times \mathbb{N} \to \mathbb{N} \times \mathbb{N}$ defined by

$$\pi(i,j) = \begin{cases} \phi(i,j), & \text{if } (i,j) \in K \\ \psi(i,j), & \text{otherwise.} \end{cases}$$

is a permutation on $\mathbb{N} \times \mathbb{N}$.

But $(x_{\pi(i,j)}) \notin {}_2BV_\sigma^I(f)$ and hence $(x_{\pi(i,j)}) \notin {}_2({}_0BV_\sigma^I(f))$ showing that ${}_2BV_\sigma^I(f)$ and ${}_2({}_0BV_\sigma^I(f))$ are not symmetric double sequence spaces.

5. Conclusion

In this chapter, we study different forms of BV_σ double sequence spaces of invariant means with the help of ideal, operators and some functions such as Orlicz function and modulus function. The chapter shows the potential of the new theoretical tools to deal with the convergence problems of sequences in sigma bounded variation, occurring in many branches of science, engineering and applied mathematics.

Acknowledgements

The authors express their gratitude to our Chairman Prof. Mohammad Imdad for his advices and continuous support. We are grateful to the learned referees for careful reading of our manuscript and considering the same for publication.

Conflict of interest

The authors declare that they have no competing interests.

Author details

Vakeel Ahmad Khan*, Hira Fatima and Mobeen Ahmad
Department of Mathematics, Aligarh Muslim University, Aligarh, India

*Address all correspondence to: vakhanmaths@gmail.com

References

[1] Fast H. Sur la convergence statistique. Colloquium Mathematicum; **1951**(2):241-244

[2] Schoenberg IJ. The integrability of certain functions and related summability methods. The American Mathematical Monthly. 1959;**66**:361-375

[3] Kostyrko P, Salat T, Wilczynski W. I-convergence. Real Analysis Exchange. 2000;**26**(2):669-686

[4] Das P, Kostyrko P, Wilczynski W, Malik P. I and I^*-convergence of double sequences. Mathematica Slovaca. 2008; **58**:605-620

[5] Basarir M, Solanacan O. On some double sequence spaces. Journal of Indian Academy of Mathematical Society, Japan. 1999;**21**(2):193-200

[6] Bromwich TJI. An Introduction to the Theory of Infinite Series. New York: MacMillan Co. Ltd; 1965

[7] Habil ED. Double sequences and double series. The Islamic University Journal, Series of Natural Studies and Engineering. 2006;**14**:1-33

[8] Khan VA, Fatima H, Abdullaha SAA, Khan MD. On a new BV_σ I-convergent double sequence spaces. Theory and Application of Mathematics and Computer Science. 2016;**6**(2):187-197

[9] Mursaleen M. On some new invariant matrix methods of summability. The Quarterly Journal of Mathematics. 1983;**34**(133):77-86

[10] Raimi RA. Invariants means and invariant matrix methods of summability. Duke Mathematical Journal. 1963;**30**:81-94

[11] Banach S. Theorie des operations lineaires, Instytut Matematyczny PAN, Warszawa; 1932

[12] Lorentz GG. A contribution to the theory of divergent series. Acta Mathematica. 1948;**80**:167-190

[13] Et M. On some new Orlicz spaces. Journal of Analysis. 2001;**9**:21-28

[14] Parshar SD, Choudhary B. Sequence spaces defined by Orlicz function. Indian Journal of Pure and Applied Mathematics. 1994;**25**:419-428

[15] Maddox IJ. Sequence spaces defined by a modulus. Mathematical Proceedings of the Cambridge Philosophical Society. 1986;**100**:161-166

[16] Maddox IJ. Elements of Functional Analysis. United Kingdom: Cambridge University Press; 1970

[17] Nakano H. Modular sequence spaces. Proceedings of Japan Academy of Series A Mathematical Sciences. 1951; **27**:508-512

[18] Tripathy BC, Hazarika B. Some I-Convergent sequence spaces defined by Orlicz function. Acta Mathematicae Applicatae Sinica. 2011;**27**(1):149-154

[19] Khan VA, Fatima H, Abdullaha SAA, Alshlool KMAS. On paranorm BV_σ I-convergent double sequence spaces defined by an Orlicz function. Analysis. 2017;**37**(3):157-167

[20] Khan VA, Khan N. On a new I-convergent double sequence space. Hindawi Publication Corporation International Journal of Analysis. 2013; **2013**:7. Article ID 126163

Chapter 7

Simple Approach to Special Polynomials: Laguerre, Hermite, Legendre, Tchebycheff, and Gegenbauer

Vicente Aboites and Miguel Ramírez

Abstract

Special polynomials: Laguerre, Hermite, Legendre, Tchebycheff and Gegenbauer are obtained through well-known linear algebra methods based on Sturm-Liouville theory. A matrix corresponding to the differential operator is found and its eigenvalues are obtained. The elements of the eigenvectors obtained correspond to each mentioned polynomial. This method contrasts in simplicity with standard methods based on solving the differential equation by means of power series, obtaining them through a generating function, using the Rodrigues formula for each polynomial, or by means of a contour integral.

Keywords: special polynomials, special functions, linear algebra, eigenvalues, eigenvectors

1. Introduction

The polynomials covered in this chapter are solutions to an ordinary differential equation (ODE), the hypergeometric equation. In general, the hypergeometric equation may be written as:

$$s(x)F''(x) + t(x)F'(x) + \lambda F(x) = 0, \tag{1}$$

where $F(x)$ is a real function of a real variable $F : U \to \mathbb{R}$, where $U \subset \mathbb{R}$ is an open subset of the real line, and $\lambda \in \mathbb{R}$ a corresponding eigenvalue, and the functions $s(x)$ and $t(x)$ are real polynomials of at most second order and first order, respectively.

There are different cases obtained, depending on the kind of the $s(x)$ function in Eq. (1). When $s(x)$ is a constant, Eq. (1) takes the form $F''(x) - 2\alpha xF'(x) + \lambda F(x) = 0$, and if $\alpha = 1$ one obtains the Hermite polynomials. When $s(x)$ is a polynomial of the first degree, Eq. (1) takes the form $xF''(x) + (-\alpha x + \beta + 1)F'(x) + \lambda F(x) = 0$, and when $\alpha = 1$ and $\beta = 0$, one obtains the Laguerre polynomials. There are three different cases when $s(x)$ is a polynomial of the second degree. When the second degree polynomial has two different real roots, Eq. (1) takes the form $(1 - x^2)F''(x) + [\beta - \alpha - (\alpha + \beta + 2)x]F'(x) + \lambda F(x) = 0$; this is the Jacobi equation, and for different values of α and β, one obtains particular cases of polynomials: Gegenbauer polynomials if $\alpha = \beta$, Tchebycheff I and II if $\alpha = \beta = \pm 1/2$, and Legendre polynomials if $\alpha = \beta = 0$. When the second degree polynomial has one double root, Eq. (1) takes the form $x^2F''(x) + [(\alpha + 2)x + \beta]F'(x) + \lambda F(x) = 0$, and when $\alpha = -1$ and $\beta = 0$, one obtains the Bessel polynomials. Finally, when the second degree polynomial has two complex roots, Eq. (1) takes the form $(1 + x)^2F''(x) + (2\beta x + \alpha)F'(x) + \lambda F(x) = 0$, which is the Romanovski equation [1]. These results are summarized in **Table 1**.

The Sturm-Liouville Theory is covered in most advanced physics and engineering courses. In this context, an eigenvalue equation sometimes takes the more general self-adjoint form: $\mathcal{L}u(x) + \lambda w(x)u(x) = 0$, where $\mathcal{L}$ is a differential operator; $\mathcal{L}u(x) = \frac{d}{dx}\left[p(x)\frac{du(x)}{dx}\right] + q(x)u(x)$, λ an eigenvalue, and $w(x)$ is known as a weight or density function. The analysis of this equation and its solutions is called the Sturm-Liouville theory. Specific forms of $p(x)$, $q(x)$, λ and $w(x)$ are given for Legendre, Laguerre, Hermite and other well-known equations in the given references. There, the close analogy of this theory with linear algebra concepts is also shown. For example, functions here take the role of vectors there, and linear operators here take that of matrices there. Finally, the diagonalization of a real symmetric matrix corresponds to the solution of an ordinary differential equation, defined by a self-adjoint operator $\mathcal{L}$, in terms of its eigenfunctions, which are the "continuous" analog of the eigenvectors [2, 3].

$s(x)$	Canonical form and weight function		Example
Constant	$F''(x) - 2\alpha xF'(x) + \lambda F(x) = 0$ $w(x) = e^{-\alpha x^2}$	(2) (3)	When $\alpha = 1$ one obtains the Hermite equation, $F(x) = H(x)$; this produces the Hermite polynomials, denoted as $\{H_n^{(\alpha)}\}$.
First degree	$xF''(x) + (-\alpha x + \beta + 1)F'(x) + \lambda F(x) = 0$ $w(x) = x^\beta e^{-\alpha x}$	(4) (5)	When $\alpha = 1$ and $\beta = 0$, one obtains the Laguerre equation, $F(x) = L(x)$; this produces the Laguerre polynomials, denoted as $\{L_n^{(\alpha,\beta)}\}$.
Second degree: with two different real roots	$(1 - x^2)F''(x) + [\beta - \alpha - (\alpha + \beta + 2)x]F'(x) + \lambda F(x) = 0$ $w^{(\alpha,\beta)}(x) = (1 - x)^\alpha(1 + x)^\beta$	(6) (7)	Eq. (6) is the Jacobi equation, considering $F(x) = P(x)$, and for each pair (α, β), one obtains the Jacobi polynomials, denoted as $\{P_n^{(\alpha,\beta)}\}$. Particular cases: Gegenbauer polynomials if $\alpha = \beta$, Tchebycheff I and II if $\alpha = \beta = \pm\frac{1}{2}$, and Legendre polynomials if $\alpha = \beta = 0$.
Second degree: with one double real root	$x^2F''(x) + [(\alpha + 2)x + \beta]F'(x) + \lambda F(x) = 0$ $w^{(\alpha,\beta)}(x) = x^\alpha e^{-\frac{\beta}{x}}$	(8) (9)	When $\alpha = -1$ and $\beta = 0$, one obtains the Bessel equation, $F(x) = B(x)$; this produces the Bessel polynomials, denoted as $\{B_n^{(\alpha,\beta)}\}$.
Second degree: with two complex roots	$(1 + x)^2F''(x) + (2\beta x + \alpha)F'(x) + \lambda F(x) = 0$ $w^{(\alpha,\beta)}(x) = (1 + x^2)^{\beta-1}e^{-\alpha\cot^{-1}x}$	(10) (11)	Eq. (10) is the Romanovski equation; considering $F(x) = R(x)$, then one obtains the Romanovski polynomials, denoted as $\{R_n^{(\alpha,\beta)}\}$.

Table 1.
Polynomials obtained depending on the $s(x)$ function of Eq. (1).

The next section shows some of the most important applications of Hermite, Gegenbauer, Tchebycheff, Laguerre and Legendre polynomials in applied Mathematics and Physics. These polynomials are of great importance in mathematical physics, the theory of approximation, the theory of mechanical quadrature, engineering, and so forth.

2. Physical applications

2.1 Laguerre

Laguerre polynomials were named after Edmond Laguerre (1834–1886). Laguerre studied a special case in 1897, and in 1880, Nikolay Yakovlevich Sonin worked on the general case known as Sonine polynomials, but they were anticipated by Robert Murphy (1833).

The Laguerre differential equation and its solutions, that is, Laguerre polynomials, are found in many important physical problems, such as in the description of the transversal profile of Laguerre-Gaussian laser beams [4]. The practical importance of Laguerre polynomials was enhanced by Schrödinger's wave mechanics, where they occur in the radial wave functions of the hydrogen atom [5].

The most important single application of the Laguerre polynomials is in the solution of the Schrödinger wave equation for the hydrogen atom. This equation is

$$-\frac{\hbar^2}{2m}\nabla^2\psi - \frac{Ze^2}{r}\psi = E\psi, \tag{12}$$

in which $Z = 1$ for hydrogen, 2 for single ionized helium, and so on. Separating variables, we find that the angular dependence of ψ is $Y_L^M(\theta, \varphi)$. The radial part, $R(r)$, satisfies the equation

$$-\frac{\hbar^2}{2m}\frac{1}{r^2}\frac{d}{dr}\left(r^2\frac{dR}{dr}\right) - \frac{Ze^2}{r}R + \frac{L(L+1)}{r^2}R = ER. \tag{13}$$

By use of the abbreviations

$$\rho = \alpha r, \quad \text{with } \alpha^2 = -\frac{8mE}{\hbar^2}, \ E < 0, \ \lambda = \frac{2mZe^2}{\alpha\hbar^2}, \tag{14}$$

Eq. (14) becomes

$$\frac{1}{\rho^2}\frac{d}{d\rho}\left(\rho^2\frac{d\chi(\rho)}{d\rho}\right) + \left(\frac{\lambda}{\rho} - \frac{1}{4} - \frac{L(L+1)}{\rho^2}\right)\chi(\rho) = 0, \tag{15}$$

where $\chi(\rho) = R(\rho/\alpha)$. Eq. (15) is satisfied by

$$\rho\chi(\rho) = e^{-\frac{\rho}{2}}\rho^{L+1}L_{\lambda-L-1}^{2L+1}(\rho), \tag{16}$$

in which k is replaced by $2L + 1$ and n by $\lambda - L - 1$, in order to consider the associated Laguerre polynomials $L_n^k(\rho)$.

These polynomials are also used in problems involving the integration of Helmholtz's equation in parabolic coordinates, in the theory of propagation of electromagnetic waves along transmission lines, in describing the static Wigner functions of oscillator systems in quantum mechanics in phase space [6], etc.

2.2 Hermite

Hermite polynomials were defined into the theory of probability by Pierre-Simon Laplace in 1810, and Charles Hermite extended them to include several variables and named them in 1864 [7].

Hermite polynomials are used to describe the transversal profile of Hermite-Gaussian laser beams [4], but mainly to analyze the quantum mechanical simple harmonic oscillator [8]. For a potential energy $V = \frac{1}{2}Kz^2 = \frac{1}{2}m\omega^2 z^2$ (force $F = \nabla V = -Kz$), the Schrödinger wave equation is

$$-\frac{\hbar^2}{2m}\nabla^2\Psi(z) + \frac{1}{2}Kz^2\Psi(z) = E\Psi(z). \tag{17}$$

The oscillating particle has mass m and total energy E. By use of the abbreviations

$$x = \alpha z \text{ with } \alpha^4 = \frac{mK}{\hbar^2} = \frac{m^2\omega^2}{\hbar^2}, \lambda = \frac{2E}{\hbar}\left(\frac{m}{K}\right)^{1/2} = \frac{2E}{\hbar\omega}, \tag{18}$$

in which ω is the angular frequency of the corresponding classical oscillator, Eq. (17) becomes

$$\frac{d^2\psi(x)}{dx^2} + (\lambda - x^2)\psi(x) = 0, \tag{19}$$

where $\psi(x) = \Psi(z) = \Psi(x/\alpha)$. With $\lambda = 2n+1$, Eq. (19) is satisfied by

$$\psi_n(x) = 2^{-\frac{n}{2}}\pi^{-\frac{1}{4}}(n!)^{-\frac{1}{2}}e^{-\frac{x^2}{2}}H_n(x), \tag{20}$$

where $H_n(x)$ corresponds to Hermite polynomials.

Hermite polynomials also appear in probability as the Edgeworth series, in combinatorics as an example of an Appell sequence, obeying the umbral calculus, in numerical analysis as Gaussian quadrature, etc.

2.3 Legendre

Legendre polynomials were first introduced in 1782 by Adrien-Marie Legendre. Spherical harmonics are an important class of special functions that are closely related to these polynomials. They arise, for instance, when Laplace's equation is solved in spherical coordinates. Since continuous solutions of Laplace's equation are *harmonic functions*, these solutions are called *spherical harmonics* [9].

In the separation of variables of Laplace's equation, Helmholtz's or the space-dependence of the classical wave equation, and the Schrödinger wave equation for central force fields,

$$\nabla^2\psi + k^2 f(r)\psi = 0, \tag{21}$$

the angular dependence, coming entirely from the Laplacian operator, is

$$\frac{\Phi(\phi)}{\sin(\theta)}\frac{d}{d\theta}\left(\sin\theta\frac{d\Theta}{d\theta}\right) + \frac{\Theta(\theta)}{\sin^2\theta}\frac{d^2\Phi(\phi)}{d\phi^2} + n(n+1)\Theta(\theta)\Phi(\phi) = 0. \tag{22}$$

The separated azimuthal equation is

$$\frac{1}{\Phi(\phi)}\frac{d^2\Phi(\phi)}{d\phi^2} = -m^2, \tag{23}$$

with an orthogonal and normalized solution,

$$\Phi_m = \frac{1}{\sqrt{2\pi}}e^{im\phi}. \tag{24}$$

Splitting off the azimuthal dependence, the polar angle dependence (θ) leads to the associated Legendre equation, which is satisfied by the associated Legendre functions; that is, $\Theta(\theta) = P_n^m(cos\theta)$. Normalizing the associated Legendre function, one obtains the orthonormal function

$$\wp_n^m(cos\theta) = \sqrt{\frac{2n+1}{2}\frac{(n-m)!}{(n+m)!}}P_n^m(cos\theta). \tag{25}$$

Taking the product of Eqs. (24) and (25) to define,

$$Y_n^m(\theta,\phi) \equiv (-1)^m\sqrt{\frac{2n+1}{4\pi}\frac{(n-m)!}{(n+m)!}}P_n^m(cos\theta)e^{im\phi}. \tag{26}$$

These $Y_n^m(\theta,\phi)$ are the spherical harmonics [10].

Legendre polynomials are frequently encountered in physics and other technical fields. Some examples are the coefficients in the expansion of the Newtonian potential that gives the gravitational potential associated to a point mass or the Coulomb potential associated to a point charge, the gravitational and electrostatic potential inside a spherical shell, steady-state heat conduction problems in spherical problems inside a homogeneous solid sphere, and so forth [11].

2.4 Tchebycheff

Tchebycheff polynomials, named after Pafnuty Tchebycheff (also written as Chebyshev, Tchebyshev or Tschebyschow), are important in approximation theory because the roots of the Tchebycheff polynomials of the first kind, which are also called Tchebycheff nodes, are used as nodes in polynomial interpolation. Approximation theory is concerned with how functions can best be approximated with simpler functions, and through quantitatively characterizing the errors introduced thereby.

One can obtain polynomials very close to the optimal one by expanding the given function in terms of Tchebycheff polynomials, and then cutting off the expansion at the desired degree. This is similar to the Fourier analysis of the function, using the Tchebycheff polynomials instead of the usual trigonometric functions.

If one calculates the coefficients in the Tchebycheff expansion for a function,

$$f(x) \sim \sum_{i=0}^{\infty} c_i T_i(x), \tag{27}$$

and then cuts off the series after the T_N term, one gets an Nth-degree polynomial approximating $f(x)$.

Tchebycheff polynomials are also found in many important physics, mathematics and engineering problems. A capacitor whose plates are two eccentric spheres is an interesting example [12], another one can be found in aircraft aerodynamics [13], etc.

2.5 Gegenbauer

Gegenbauer polynomials, named after Leopold Gegenbauer, and often called ultraspherical polynomials, include Legendre and Tchebycheff polynomials as special or limiting cases, and at the same time, Gegenbauer polynomials are a special case of Jacobi polynomials (see **Table 1**).

Gegenbauer polynomials appear naturally as extensions of Legendre polynomials in the context of potential theory and harmonic analysis. They also appear in the theory of Positive-definite functions [14].

Since Gegenbauer polynomials are a general case of Legendre and Tchebycheff polynomials, more applications are shown in Section 2.3 and 2.4.

The most common methods to obtain the special polynomials are described in the next section.

3. Special polynomials

To obtain the polynomials described in the previous section, one can use different methods, some tougher than others. These polynomials are typically obtained as a result of the solution of each specific differential equation by means of the power series method. Usually, it is also shown that they can be obtained through a generating function and also by using the Rodrigues formula for each special polynomial, or finally, through a contour integral. Most Mathematical Methods courses also include a study of the properties of these polynomials, such as orthogonality, completeness, recursion relations, special values, asymptotic expansions and their relation to other functions, such as polynomials and hypergeometric functions. There is no doubt that this is a challenging and demanding subject that requires a great deal of attention from most students.

3.1 Differential equation

The most common way to solve the special polynomials is solving the associated differential equation through power series and the Frobenius method $y = \sum_{n=0}^{\infty} a_n x^n$. The corresponding polynomials satisfy the following differential equations:

the Laguerre differential equation,

$$xy'' + (1 - x)y' + ny = 0, \tag{28}$$

the Hermite differential equation,

$$y' - 2xy' + 2ny = 0, \tag{29}$$

the Legendre differential equation,

$$\left(1 - x^2\right)y'' - 2xy' + n(n+1)y = 0\,, \tag{30}$$

the Tchebycheff differential equation,

$$\left(1 - x^2\right)y'' - xy' + n^2 y = 0, \tag{31}$$

and the Gegenbauer differential equation,

$$\left(1 - x^2\right)y'' - (2\lambda + 1)xy' + n(n + 2\lambda)y = 0, \tag{32}$$

with $n = 0, 1, 2, 3, \ldots$ in all the previous cases. Note that if $\lambda = \frac{1}{2}$, Eq. (32) reduces to the Legendre differential equation (Eq. (30)), and if $\lambda = 0$, Eq. (32) reduces to the Tchebycheff differential equation (Eq. (31)).

3.2 Rodrigues formula

For polynomials $\psi_n(x)$, with interval I, weight function $w(x)$, and an eigenvalue equation of the form

$$p(x)\psi_n''(x) + q(x)\psi_n'(x) + \lambda_n\psi_n(x) = 0, \tag{33}$$

and with $q(x) = \frac{(p(x)w(x))'}{w(x)}$, the general formula

$$\psi_n(x) = w(x)^{-1}\frac{d^n}{dx^n}[p(x)^n w(x)] \tag{34}$$

is known as the *Rodrigues formula,* useful to obtain the nth-degree polynomial of ψ [15].

3.3 Generating function and contour integral

Let Γ be a curve that encloses $x \in I$ but excludes the endpoints of I. Then, considering the Cauchy integral formula [16] for derivatives of $w(x)p(x)^n$ to derive an integral formula from Eq. (34), one obtains

$$\frac{\psi_n(x)}{n!} = \frac{1}{2\pi i}\int_\Gamma \frac{w(z)}{w(x)}\frac{p(z)^n}{(z-x)^n}\frac{dz}{z-x}. \tag{35}$$

The *generating function* for the orthogonal polynomials $\left\{\frac{\psi_n(x)}{n!}\right\}$ is defined as

$$G(x,s) = \sum_{n=0}^{\infty}\frac{\psi_n(x)}{n!}s^n. \tag{36}$$

In the following section, Laguerre [2], Hermite [17], Legendre, Tchebycheff [18] and Gegenbauer [3] polynomials are obtained through a simple method, using basic linear algebra concepts, such as the eigenvalue and the eigenvector of a matrix.

4. Simple approach to special polynomials

The general algebraic polynomial of degree n,

$$a_0 + a_1x + a_2x^2 + a_3x^3 + \ldots a_nx^n, \tag{37}$$

with $a_o, a_1, \ldots, a_n \in \Re$, is represented by vector

$$A_n = \begin{bmatrix} a_0 \\ a_1 \\ a_2 \\ a_3 \\ \vdots \\ a_n \end{bmatrix}. \tag{38}$$

Taking the first derivative of the above polynomial (x), one obtains the polynomial

$$\frac{d}{dx}\left[a_0 + a_1x + a_2x^2 + a_3x^3 + \ldots\ a_nx^n\right] = a_1 + 2a_2x + 3a_3x^2 + \ldots\ na_nx^{n-1}, \tag{39}$$

which may be written as

$$\frac{dA_n}{dx} = \begin{bmatrix} a_1 \\ 2a_2 \\ 3a_3 \\ \vdots \\ na_n \\ 0 \end{bmatrix}. \tag{40}$$

Taking the second derivative of the polynomial (Eq. (37)) one obtains

$$\frac{d^2}{dx^2}\left[a_0 + a_1x + a_2x^2 + a_3x^3 + \ldots\ a_nx^n\right] = 2a_2 + 6a_3x + \ldots\ n(n-1)a_nx^{n-2}, \tag{41}$$

which may be written as

$$\frac{d^2A_n}{dx^2} = \begin{bmatrix} 2a_2 \\ 6a_3 \\ \vdots \\ n(n-1)a_n \\ 0 \\ 0 \end{bmatrix}. \tag{42}$$

Using Eq. (40), Eq. (39) may be written as

$$\begin{bmatrix} 0 & 1 & 0 & 0 & \cdots & 0 \\ 0 & 0 & 2 & 0 & \cdots & 0 \\ 0 & 0 & 0 & 3 & \cdots & 0 \\ \vdots & \vdots & \vdots & \vdots & \ddots & \vdots \\ 0 & 0 & 0 & 0 & \cdots & n \\ 0 & 0 & 0 & 0 & \cdots & 0 \end{bmatrix} \begin{bmatrix} a_0 \\ a_1 \\ a_2 \\ a_3 \\ \vdots \\ a_n \end{bmatrix} = \begin{bmatrix} a_1 \\ 2a_2 \\ 3a_3 \\ \vdots \\ na_n \\ 0 \end{bmatrix}; \tag{43}$$

therefore, the first derivative operator A_n may be written as

$$\frac{d}{dx} \rightarrow \begin{bmatrix} 0 & 1 & 0 & 0 & \cdots & 0 \\ 0 & 0 & 2 & 0 & \cdots & 0 \\ 0 & 0 & 0 & 3 & \cdots & 0 \\ \vdots & \vdots & \vdots & \vdots & \ddots & \vdots \\ 0 & 0 & 0 & 0 & \cdots & n \\ 0 & 0 & 0 & 0 & \cdots & 0 \end{bmatrix}. \tag{44}$$

Doing the same for Eq. (41),

$$\begin{bmatrix} 0 & 0 & 2 & 0 & \cdots & 0 \\ 0 & 0 & 0 & 6 & \cdots & 0 \\ \vdots & \vdots & \vdots & \vdots & \ddots & \vdots \\ 0 & 0 & 0 & 0 & \cdots & n(n-1) \\ 0 & 0 & 0 & 0 & \cdots & 0 \\ 0 & 0 & 0 & 0 & \cdots & 0 \end{bmatrix} \begin{bmatrix} a_0 \\ a_1 \\ a_2 \\ \vdots \\ a_{n-1} \\ a_n \end{bmatrix} = \begin{bmatrix} a_1 \\ 2a_2 \\ \vdots \\ n(n-1)a_n \\ 0 \\ 0 \end{bmatrix}, \tag{45}$$

the second derivative operator A_n may be written as

$$\frac{d^2}{dx^2} \rightarrow \begin{bmatrix} 0 & 0 & 2 & 0 & \cdots & 0 \\ 0 & 0 & 0 & 6 & \cdots & 0 \\ \vdots & \vdots & \vdots & \vdots & \ddots & \vdots \\ 0 & 0 & 0 & 0 & \cdots & n(n-1) \\ 0 & 0 & 0 & 0 & \cdots & 0 \\ 0 & 0 & 0 & 0 & \cdots & 0 \end{bmatrix}. \tag{46}$$

4.1 Laguerre

The Laguerre differential operator is given by.

$$x\frac{d^2}{dx^2} + (1-x)\frac{d}{dx}; \tag{47}$$

substituting Eqs. (41) and (44) into Eq. (47),

$$x\left[2a_2 + 6a_3x + \ldots + n(n-1)a_nx^{n-2}\right] + (1-x)\left[a_1 + 2a_2x + 3a_3x^2 + \ldots + na_nx^{n-1}\right] \\ = a_1 + (4a_2 - a_1)x + (9a_3 - 2a_2)x^2 + (16a_4 + 3a_3)x^3 + \cdots - na_n, \tag{48}$$

which may be written as

$$\begin{bmatrix} 0 & 1 & 0 & 0 & 0 & \cdots & 0 \\ 0 & -1 & 4 & 0 & 0 & \cdots & 0 \\ 0 & 0 & -2 & 9 & 0 & \cdots & 0 \\ 0 & 0 & 0 & -3 & 16 & \cdots & 0 \\ \vdots & \vdots & \vdots & \vdots & \vdots & \ddots & \vdots \\ 0 & 0 & 0 & 0 & 0 & \cdots & -n \end{bmatrix} \begin{bmatrix} a_0 \\ a_1 \\ a_2 \\ a_3 \\ \vdots \\ a_n \end{bmatrix} = \begin{bmatrix} a_1 \\ 4a_2 - a_1 \\ 9a_3 - 2a_2 \\ 16a_4 - 3a_3 \\ \vdots \\ -na_n \end{bmatrix}. \tag{49}$$

For simplicity, the Laguerre differential operator, as a $4x4$ matrix, is represented by

$$x\frac{d^2}{dx^2} + (1-x)\frac{d}{dx} \rightarrow \begin{bmatrix} 0 & 1 & 0 & 0 \\ 0 & -1 & 4 & 0 \\ 0 & 0 & -2 & 9 \\ 0 & 0 & 0 & -3 \end{bmatrix}. \tag{50}$$

The eigenvalues of a matrix M are the values that satisfy the equation $Det(M - \lambda I) = 0$. However, since the matrix (Eq. (50)) is a triangular matrix, the eigenvalues λ_i of this matrix are the elements of the diagonal, namely: $\lambda_1 = 0$, $\lambda_2 = -1$, $\lambda_3 = -2$, $\lambda_4 = -3$. The corresponding eigenvectors are the solutions of the equation $(M - \lambda_i I) \cdot v = 0$, where the eigenvector $v = [a_0, a_1, a_2, a_3]^T$:

$$\begin{bmatrix} 0-\lambda_i & 1 & 0 & 0 \\ 0 & -1-\lambda_i & 4 & 0 \\ 0 & 0 & -2-\lambda_i & 9 \\ 0 & 0 & 0 & -3-\lambda_i \end{bmatrix} \begin{bmatrix} a_0 \\ a_1 \\ a_2 \\ a_3 \end{bmatrix} = \begin{bmatrix} 0 \\ 0 \\ 0 \\ 0 \end{bmatrix}. \tag{51}$$

Substituting eigenvalue $\lambda_1 = 0$ in Eq. (51), we obtain eigenvector v_1:

$$v_1 = \begin{bmatrix} 1 \\ 0 \\ 0 \\ 0 \end{bmatrix}; \tag{52}$$

the elements of this eigenvector correspond to the first Laguerre polynomial, $L_0(x) = 1$.

Substituting eigenvalue $\lambda_2 = -1$ in Eq. (51), we obtain eigenvector v_2:

$$v_2 = \begin{bmatrix} 1 \\ -1 \\ 0 \\ 0 \end{bmatrix}; \tag{53}$$

the elements of this eigenvector correspond to the second Laguerre polynomial, $L_1(x) = 1 - x$.

Substituting eigenvalue $\lambda_3 = -2$ in Eq. (51), we obtain eigenvector v_3:

$$v_3 = \begin{bmatrix} 1 \\ -2 \\ \frac{1}{2} \\ 0 \end{bmatrix}; \tag{54}$$

the elements of this eigenvector correspond to the third Laguerre polynomial, $L_2(x) = 1 - 2x + \frac{1}{2}x^2$.

Substituting eigenvalue $\lambda_4 = -3$ in Eq. (51), we obtain eigenvector v_4:

$$v_4 = \begin{bmatrix} 1 \\ -3 \\ \frac{3}{2} \\ -\frac{1}{6} \end{bmatrix}; \tag{55}$$

the elements of this eigenvector correspond to the fourth Laguerre polynomial, $L_3(x) = 1 - 3x + \frac{3}{2}x^2 - \frac{1}{6}x^3$.

4.2 Hermite

The Hermite differential operator is given by

$$\frac{d^2}{dx^2} - 2x\frac{d}{dx}; \tag{56}$$

substituting Eqs. (41) and (44) into Eq. (56),

$$\begin{aligned}&\left[2a_2 + 6a_3x + \ldots + n(n-1)a_nx^{n-2}\right] - 2x\left[a_1 + 2a_2x + 3a_3x^2 + \ldots + na_nx^{n-1}\right]\\&= 2a_2 + (6a_3 - 2a_1)x + (12a_4 - 4a_2)x^2 + (20a_5 - 6a_3)x^3 + \cdots - 2na_n,\end{aligned} \tag{57}$$

which may be written as

$$\begin{bmatrix} 0 & 0 & 2 & 0 & 0 & \cdots & 0 \\ 0 & -2 & 0 & 6 & 0 & \cdots & 0 \\ 0 & 0 & -4 & 0 & 12 & \cdots & 0 \\ 0 & 0 & 0 & -6 & 0 & \cdots & 0 \\ \vdots & \vdots & \vdots & \vdots & \vdots & \ddots & \vdots \\ 0 & 0 & 0 & 0 & 0 & \cdots & -2n \end{bmatrix} \begin{bmatrix} a_0 \\ a_1 \\ a_2 \\ a_3 \\ \vdots \\ a_n \end{bmatrix} = \begin{bmatrix} 2a_2 \\ 6a_3 - 2a_1 \\ 12a_4 - 4a_2 \\ 20a_5 - 6a_3 \\ \vdots \\ -2na_n \end{bmatrix}. \tag{58}$$

For simplicity, the Hermite differential operator, as a $4x4$ matrix, is represented by

$$\frac{d^2}{dx^2} - 2x\frac{d}{dx} \rightarrow \begin{bmatrix} 0 & 0 & 2 & 0 \\ 0 & -2 & 0 & 6 \\ 0 & 0 & -4 & 0 \\ 0 & 0 & 0 & -6 \end{bmatrix}. \tag{59}$$

The eigenvalues of a matrix M are the values that satisfy the equation $Det(M - \lambda I) = 0$. However, since the matrix (Eq. (59)) is a triangular matrix, the eigenvalues λ_i of this matrix are the elements of the diagonal, namely: $\lambda_1 = 0$, $\lambda_2 = -2$, $\lambda_3 = -4$, $\lambda_4 = -6$. The corresponding eigenvectors are the solutions of the equation $(M - \lambda_i I) \cdot v = 0$, where the eigenvector $v = [a_0, a_1, a_2, a_3]^T$:

$$\begin{bmatrix} 0 - \lambda_i & 0 & 2 & 0 \\ 0 & -2 - \lambda_i & 0 & 6 \\ 0 & 0 & -4 - \lambda_i & 0 \\ 0 & 0 & 0 & -6 - \lambda_i \end{bmatrix} \begin{bmatrix} a_0 \\ a_1 \\ a_2 \\ a_3 \end{bmatrix} = \begin{bmatrix} 0 \\ 0 \\ 0 \\ 0 \end{bmatrix}. \tag{60}$$

Substituting eigenvalue $\lambda_1 = 0$ in Eq. (60), we obtain eigenvector v_1:

$$v_1 = \begin{bmatrix} 1 \\ 0 \\ 0 \\ 0 \end{bmatrix}; \tag{61}$$

the elements of this eigenvector correspond to the first Hermite polynomial, $H_0(x) = 1$.

Substituting eigenvalue $\lambda_2 = -2$ in Eq. (60), we obtain eigenvector v_2:

$$v_2 = \begin{bmatrix} 0 \\ 2 \\ 0 \\ 0 \end{bmatrix}; \tag{62}$$

the elements of this eigenvector correspond to the second Hermite polynomial, $H_1(x) = 2x$.

Substituting eigenvalue $\lambda_3 = -4$ in Eq. (60), we obtain eigenvector v_3:

$$v_3 = \begin{bmatrix} -2 \\ 0 \\ 4 \\ 0 \end{bmatrix}; \tag{63}$$

the elements of this eigenvector correspond to the third Hermite polynomial, $H_2(x) = 4x^2 - 2$.

Substituting eigenvalue $\lambda_4 = -6$ in Eq. (60), we obtain eigenvector v_4:

$$v_4 = \begin{bmatrix} 0 \\ -12 \\ 0 \\ 8 \end{bmatrix}; \tag{64}$$

the elements of this eigenvector correspond to the fourth Hermite polynomial, $H_3(x) = 8x^3 - 12x$.

4.3 Legendre

The Legendre differential operator is given by

$$\left(1 - x^2\right)\frac{d^2}{dx^2} - 2x\frac{d}{dx}; \tag{65}$$

substituting Eqs. (41) and (44) into Eq. (65),

$$\begin{aligned}\left(1 - x^2\right)\left[2a_2 + 6a_3x + \dots + n(n-1)a_nx^{n-2}\right] - 2x\left[a_1 + 2a_2x + 3a_3x^2 + \dots + na_nx^{n-1}\right] \\ = 2a_2 + (6a_3 - 2a_1)x + (12a_4 - 6a_2)x^2 + (20a_5 - 12a_3)x^3 + \cdots - \left(n^2 + n\right)a_n,\end{aligned} \tag{66}$$

which may be written as

$$\begin{bmatrix} 0 & 0 & 2 & 0 & 0 & \cdots & 0 \\ 0 & -2 & 0 & 6 & 0 & \cdots & 0 \\ 0 & 0 & -6 & 0 & 12 & \cdots & 0 \\ 0 & 0 & 0 & -12 & 0 & \cdots & 0 \\ \vdots & \vdots & \vdots & \vdots & \vdots & \ddots & \vdots \\ 0 & 0 & 0 & 0 & 0 & \cdots & -\left(n^2 + n\right) \end{bmatrix} \begin{bmatrix} a_0 \\ a_1 \\ a_2 \\ a_3 \\ \vdots \\ a_n \end{bmatrix} = \begin{bmatrix} 2a_2 \\ 6a_3 - 2a_1 \\ 12a_4 - 6a_2 \\ 20a_5 - 12a_3 \\ \vdots \\ -\left(n^2 + n\right)a_n \end{bmatrix}. \tag{67}$$

For simplicity, the Legendre differential operator, as a 4x4 matrix, is represented by

$$\left(1 - x^2\right)\frac{d^2}{dx^2} - 2x\frac{d}{dx} \to \begin{bmatrix} 0 & 0 & 2 & 0 \\ 0 & -2 & 0 & 6 \\ 0 & 0 & -6 & 0 \\ 0 & 0 & 0 & -12 \end{bmatrix}. \tag{68}$$

The eigenvalues of a matrix M are the values that satisfy the equation $Det(M - \lambda I) = 0$. However, since the matrix (Eq. (68)) is a triangular matrix, the eigenvalues λ_i of this matrix are the elements of the diagonal, namely: $\lambda_1 = 0$, $\lambda_2 = -2$, $\lambda_3 = -6$, $\lambda_4 = -12$. The corresponding eigenvectors are the solutions of the equation $(M - \lambda_i I) \cdot v = 0$, where the eigenvector $v = [a_0, a_1, a_2, a_3]^T$:

$$\begin{bmatrix} 0-\lambda_i & 0 & 2 & 0 \\ 0 & -2-\lambda_i & 0 & 6 \\ 0 & 0 & -6-\lambda_i & 0 \\ 0 & 0 & 0 & -12-\lambda_i \end{bmatrix} \begin{bmatrix} a_0 \\ a_1 \\ a_2 \\ a_3 \end{bmatrix} = \begin{bmatrix} 0 \\ 0 \\ 0 \\ 0 \end{bmatrix}. \tag{69}$$

Substituting eigenvalue $\lambda_1 = 0$ in Eq. (69), we obtain eigenvector v_1:

$$v_1 = \begin{bmatrix} 1 \\ 0 \\ 0 \\ 0 \end{bmatrix}; \tag{70}$$

the elements of this eigenvector correspond to the first Legendre polynomial, $P_0(x) = 1$.

Substituting eigenvalue $\lambda_2 = -2$ in Eq. (69), we obtain eigenvector v_2:

$$v_2 = \begin{bmatrix} 0 \\ 1 \\ 0 \\ 0 \end{bmatrix}; \tag{71}$$

the elements of this eigenvector correspond to the second Legendre polynomial, $P_1(x) = x$.

Substituting eigenvalue $\lambda_3 = -6$ in Eq. (69), we obtain eigenvector v_3:

$$v_3 = \begin{bmatrix} 1 \\ 0 \\ -3 \\ 0 \end{bmatrix}; \tag{72}$$

the elements of this eigenvector correspond to the third Legendre polynomial, $P_2(x) = \frac{3}{2}x^2 - \frac{1}{2}$.

Substituting eigenvalue $\lambda_4 = -12$ in Eq. (69), we obtain eigenvector v_4:

$$v_4 = \begin{bmatrix} 0 \\ 3 \\ 0 \\ -5 \end{bmatrix}; \tag{73}$$

the elements of this eigenvector correspond to the fourth Legendre polynomial, $P_3(x) = \frac{5}{2}x^3 - \frac{3}{2}x$.

4.4 Tchebycheff

The Tchebycheff differential operator is given by

$$(1-x^2)\frac{d^2}{dx^2}-x\frac{d}{dx}; \quad (74)$$

substituting Eqs. (41) and (44) into Eq. (74),

$$\begin{aligned}(1-x^2)\left[2a_2+6a_3x+\dots+n(n-1)a_nx^{n-2}\right]-x\left[a_1+2a_2x+3a_3x^2\right.\\ \left.+\dots+na_nx^{n-1}\right]=2a_2+(6a_3-a_1)x+(12a_4-4a_2)x^2\\ +(20a_5-9a_3)x^3+\cdots-n^2a_n,\end{aligned} \quad (75)$$

which may be written as

$$\begin{bmatrix} 0 & 0 & 2 & 0 & 0 & \cdots & 0 \\ 0 & -1 & 0 & 6 & 0 & \cdots & 0 \\ 0 & 0 & -4 & 0 & 12 & \cdots & 0 \\ 0 & 0 & 0 & -9 & 0 & \cdots & 0 \\ \vdots & \vdots & \vdots & \vdots & \vdots & \ddots & \vdots \\ 0 & 0 & 0 & 0 & 0 & \cdots & -n^2 \end{bmatrix}\begin{bmatrix} a_0 \\ a_1 \\ a_2 \\ a_3 \\ \vdots \\ a_n \end{bmatrix}=\begin{bmatrix} 2a_2 \\ 6a_3-a_1 \\ 12a_4-4a_2 \\ 20a_5-9a_3 \\ \vdots \\ -n^2a_n \end{bmatrix}. \quad (76)$$

For simplicity, the Tchebycheff differential operator, as a $4x4$ matrix, is represented by

$$(1-x^2)\frac{d^2}{dx^2}-x\frac{d}{dx}\rightarrow\begin{bmatrix} 0 & 0 & 2 & 0 \\ 0 & -1 & 0 & 6 \\ 0 & 0 & -4 & 0 \\ 0 & 0 & 0 & -9 \end{bmatrix}. \quad (77)$$

The eigenvalues of a matrix M are the values that satisfy the equation $Det(M-\lambda I)=0$. However, since the matrix (Eq. (77)) is a triangular matrix, the eigenvalues λ_i of this matrix are the elements of the diagonal, namely: $\lambda_1=0$, $\lambda_2=-1$, $\lambda_3=-4$, $\lambda_4=-9$. The corresponding eigenvectors are the solutions of the equation $(M-\lambda_i I)\cdot v=0$, where the eigenvector $v=[a_0,a_1,a_2,a_3]^T$;

$$\begin{bmatrix} 0-\lambda_i & 0 & 2 & 0 \\ 0 & -1-\lambda_i & 0 & 6 \\ 0 & 0 & -4-\lambda_i & 0 \\ 0 & 0 & 0 & -9-\lambda_i \end{bmatrix}\begin{bmatrix} a_0 \\ a_1 \\ a_2 \\ a_3 \end{bmatrix}=\begin{bmatrix} 0 \\ 0 \\ 0 \\ 0 \end{bmatrix}. \quad (78)$$

Substituting eigenvalue $\lambda_1=0$ in Eq. (78), we obtain eigenvector v_1:

$$v_1=\begin{bmatrix} 1 \\ 0 \\ 0 \\ 0 \end{bmatrix}; \quad (79)$$

the elements of this eigenvector correspond to the first Tchebycheff polynomial, $T_0(x)=1$.

Substituting eigenvalue $\lambda_2=-1$ in Eq. (78), we obtain eigenvector v_2:

$$v_2 = \begin{bmatrix} 0 \\ 1 \\ 0 \\ 0 \end{bmatrix}; \tag{80}$$

the elements of this eigenvector correspond to the second Tchebycheff polynomial, $T_1(x) = x$.

Substituting eigenvalue $\lambda_3 = -4$ in Eq. (78), we obtain eigenvector v_3:

$$v_3 = \begin{bmatrix} -1 \\ 0 \\ 2 \\ 0 \end{bmatrix}; \tag{81}$$

the elements of this eigenvector correspond to the third Tchebycheff polynomial, $T_2(x) = 2x^2 - 1$.

Substituting eigenvalue $\lambda_4 = -9$ in Eq. (78), we obtain eigenvector v_4:

$$v_4 = \begin{bmatrix} 0 \\ -3 \\ 0 \\ 4 \end{bmatrix}. \tag{82}$$

the elements of this eigenvector correspond to the fourth Tchebycheff polynomial, $T_3(x) = 4x^3 - 3x$.

4.5 Gegenbauer

The Gegenbauer differential operator is given by

$$(1 - x^2)\frac{d^2}{dx^2} - (2\lambda + 1)x\frac{d}{dx}; \tag{83}$$

substituting (41) and (44) into (83),

$$\begin{aligned} &(1 - x^2)\left[2a_2 + 6a_3x + \ldots + n(n-1)a_nx^{n-2}\right] - (2\lambda + 1)x[a_1 \\ &\quad + 2a_2x + 3a_3x^2 + \ldots + na_nx^{n-1}] = 2a_2 + [6a_3 - (2\lambda + 1)a_1]x \\ &\quad + [12a_4 - 4(\lambda + 1)a_2]x^2 + [20a_5 - 3(2\lambda + 3)a_3]x^3 \\ &\quad + \cdots - \left[n^2 + 2\lambda n\right]a_n, \end{aligned} \tag{84}$$

which may be written as

$$\begin{bmatrix} 0 & 0 & 2 & 0 & 0 & \cdots & 0 \\ 0 & -(2\lambda+1) & 0 & 6 & 0 & \cdots & 0 \\ 0 & 0 & -4(\lambda+1) & 0 & 12 & \cdots & 0 \\ 0 & 0 & 0 & -3(2\lambda+3) & 0 & \cdots & 0 \\ \vdots & \vdots & \vdots & \vdots & \vdots & \ddots & \vdots \\ 0 & 0 & 0 & 0 & 0 & \cdots & -n^2 - 2\lambda n \end{bmatrix} \begin{bmatrix} a_0 \\ a_1 \\ a_2 \\ a_3 \\ \vdots \\ a_n \end{bmatrix} = \begin{bmatrix} 2a_2 \\ 6a_3 - (2\lambda+1)a_1 \\ 12a_4 - 4(\lambda+1)a_2 \\ 20a_5 - 3(2\lambda+3)a_3 \\ \vdots \\ -(n^2 + 2\lambda n)a_n \end{bmatrix}. \tag{85}$$

For simplicity, the Gegenbauer differential operator, as a 4*x*4 matrix, is represented by

$$\left(1-x^2\right)\frac{d^2}{dx^2}-(2\lambda+1)x\frac{d}{dx}\rightarrow\begin{bmatrix}0 & 0 & 2 & 0\\ 0 & -(2\lambda+1) & 0 & 6\\ 0 & 0 & -4(\lambda+1) & 0\\ 0 & 0 & 0 & -3(2\lambda+3)\end{bmatrix}. \quad (86)$$

The eigenvalues of a matrix M are the values that satisfy the equation $Det(M-\lambda' I)=0$. However, since the matrix (Eq. (86)) is a triangular matrix, the eigenvalues λ_i of this matrix are the elements of the diagonal, namely: $\lambda'_1=0$, $\lambda'_2=-(2\lambda+1)$, $\lambda'_3=-4(\lambda+1)$, $\lambda'_4=-3(2\lambda+3)$. The corresponding eigenvectors are the solutions of the equation $\left(M-\lambda'_i I\right)\cdot v=0$, where the eigenvector $v=\left[a_0,a_1,a_2,a_3\right]^T$;

$$\begin{bmatrix}0-\lambda'_i & 0 & 2 & 0\\ 0 & -(2\lambda+1)-\lambda'_i & 0 & 6\\ 0 & 0 & -4(\lambda+1)-\lambda'_i & 0\\ 0 & 0 & 0 & -3(2\lambda+3)-\lambda'_i\end{bmatrix}\cdot\begin{bmatrix}a_0\\ a_1\\ a_2\\ a_3\end{bmatrix}=\begin{bmatrix}0\\ 0\\ 0\\ 0\end{bmatrix}. \quad (87)$$

Substituting eigenvalue $\lambda'_1=0$ in Eq. (87), we obtain eigenvector v_1:

$$v_1=\begin{bmatrix}1\\ 0\\ 0\\ 0\end{bmatrix}; \quad (88)$$

the elements of this eigenvector correspond to the first Gegenbauer polynomial, $C_0^\lambda(x)=1$.

Substituting eigenvalue $\lambda'_2=-(2\lambda+1)$ in Eq. (87), we obtain eigenvector v_2:

$$v_2=\begin{bmatrix}0\\ 2\lambda\\ 0\\ 0\end{bmatrix}; \quad (89)$$

the elements of this eigenvector correspond to the second Gegenbauer polynomial, $C_1^\lambda(x)=2\lambda x$.

Substituting eigenvalue $\lambda'_3=-4(\lambda+1)$ in Eq. (87), we obtain eigenvector v_3:

$$v_3=\begin{bmatrix}-\lambda\\ 0\\ 2\lambda(1+\lambda)\\ 0\end{bmatrix}; \quad (90)$$

the elements of this eigenvector correspond to the third Gegenbauer polynomial, $C_2^\lambda(x)=-\lambda+2\lambda(1+\lambda)x^2$.

Substituting eigenvalue $\lambda'_4=-3(2\lambda+3)$ in Eq. (87), we obtain eigenvector v_4:

$$v_4 = \begin{bmatrix} 0 \\ -2\lambda(1+\lambda) \\ 0 \\ \frac{4}{3}\lambda(1+\lambda)(2+\lambda) \end{bmatrix}; \tag{91}$$

the elements of this eigenvector correspond to the fourth Gegenbauer polynomial, $C_3^{\lambda}(x) = 2\lambda(1+\lambda)x + \frac{4}{3}\lambda(1+\lambda)(2+\lambda)x^3$.

5. Conclusions

Laguerre, Hermite, Legendre, Tchebycheff and Gegenbauer polynomials are obtained in a simple and straightforward way using basic linear algebra concepts, such as the eigenvalue and the eigenvector of a matrix. Once the matrix of the corresponding differential operator is obtained, the eigenvalues of this matrix are found, and the elements of its eigenvectors correspond to the coefficients of each kind of polynomials. Using a larger matrix, higher order polynomials may be found; however, the general case for an *nxn* matrix was not obtained since it seems that in this general case, standard methods would be easier to use. The main advantage of this method lies in its easiness, since it relies on simple linear algebra concepts. This method contrasts in simplicity with standard methods based on solving the differential equation using power series, using the generating function, using the Rodrigues formula, or using a contour integral.

Acknowledgements

V. Aboites acknowledges support and useful conversations with Prof. Ernst Wintner from TU-Wien, Dr. Matei Tene from TU-Delft and Dr. Klaus Huber from Berlin. The authors acknowledge the professional English proof reading service provided by Mr. Mario Ruiz Berganza.

Author details

Vicente Aboites* and Miguel Ramírez
Centro de Investigaciones en Óptica, León, México

*Address all correspondence to: aboites@cio.mx

References

[1] Raposo A. Romanovski polynomials in selected physics problems. Central European Journal of Physics. 2007;**5**: 253-284. DOI: 10.2478/s11534-007-0018-5

[2] Aboites V. Laguerre polynomials and linear algebra. Memorias Sociedad Matemática Mexicana. 2017;**52**:3-13

[3] Ramírez M. Simple approach to Gegenbauer polynomials. International Journal of Pure and Applied Mathematics. 2018;**119**:121-129. DOI: 10.12732/ijpam.v119i1.10

[4] Siegman A. Lasers. 1st ed. California: University Science Books. p. 688

[5] Carlson B. Special Functions of Applied Mathematics. 1st ed. New York: Academic Press; 1977. p. 212

[6] Lebedev N. Special Functions and their Applications. 1st ed. New York: Dover Publications; 1972. p. 76

[7] Carlson B. Special Functions of Applied Mathematics. 1st ed. New York: Academic Press; 1977. p. 217

[8] Arfken G, Weber H. Mathematical Methods for Physicists. 4th ed. California: Academic Press; 1995. pp. 769-770

[9] Nikiforov A, Uvarov V. Special Functions of Mathematical Physics. 1st ed. Germany: Birkhäuser; 1988. p. 76

[10] Arfken G, Weber H. Mathematical Methods for Physicists. 4th ed. California: Academic Press; 1995. pp. 736-739

[11] Jackson J. Classical Electrodynamics. New York: Wiley; 1999

[12] Paszkowski S. An application of Chebyshev polynomial to a problem of electrical engineering. Journal of Computational and Applied Mathematics. 1991;**37**:5-17. DOI: 10.1016/0377-0427(91)90101-O

[13] Leng G. Compression of aircraft aerodynamic database using multivariable Chebyshev polynomials. Advances in Engineering Software. 1997;**28**:133-141. DOI: 10.1016/S0965-9978(96)00043-9

[14] Stein E, Weiss G. Introduction to Fourier Analysis on Euclidean Spaces. Nueva Jersey: Princeton University Press; 1971

[15] Beals R, Wong R. Special Functions and Orthogonal Polynomials. 1st ed. Cambridge: Cambridge University Press; 2016. pp. 94-95

[16] Beals R, Wong R. Special Functions and Orthogonal Polynomials. 1st ed. Cambridge: Cambridge University Press; 2016. p. 98

[17] Aboites V. Hermite polynomials through linear algebra. International Journal of Pure and Applied Mathematics. 2017;**114**:401-406. DOI: 10.12732/ijpam.v114i2.19

[18] Aboites V. Easy Route to Tchebycheff Polynomials. Revista Mexicana de Fisica E. 2019;**65**:12-14